CHAPTER-I

1. INTRODUCTION

A concern in the manufacturing industry today is the vibrations induced by metal cutting, e.g. turning, milling and boring operations. Turning operations and especially boring operations are associated with serious vibration-related problems. To reduce the problem of vibration and ensure that the desired shape and tolerance are achieved, extra care must be taken with production planning and in the preparations for the machining of a work-piece.

The vibration problem associated with metal cutting thus has considerable influence on important factors such as productivity, production costs, etc. A thorough investigation of the vibrations involved is therefore an important step toward solving the problem.

During an internal turning operation, the cutting tool and the boring bar are subjected to a prescribed deformation as a result of the relative motion between the tool and work-piece both in the cutting speed direction and feed direction. As a response to the prescribed deformation, the tool is subjected to traction and thermal loads on those faces that have interfacial contact with the work-piece or chip. In the metal-cutting process, during which chips are formed, the work-piece material is compressed and subjected to plastic deformation. This results in considerable strain and strain rates in the primary deformation zone (Q).

Due to the friction and the plastic deformation in the secondary and tertiary deformation zones (QII, QIII) the tool is subjected to both normal and shear traction loads on the work-piece–chip–tool interface. The traction loads vary since the deformation process in not continuous. When a material yields, the direction and the magnitude of the dislocation are determined by the state of stress as well as the structure of the material lattice. Work materials, especially iron alloys, have lattice errors. Once alloying components have been added as interstitials, the lattice errors increase and the theoretical strength decreases. However, the alloying components will prevent or lock the dislocations during plastic deformation of a material. The dislocations themselves will also prevent plastic deformation. The grain size influences the material's yield strength.

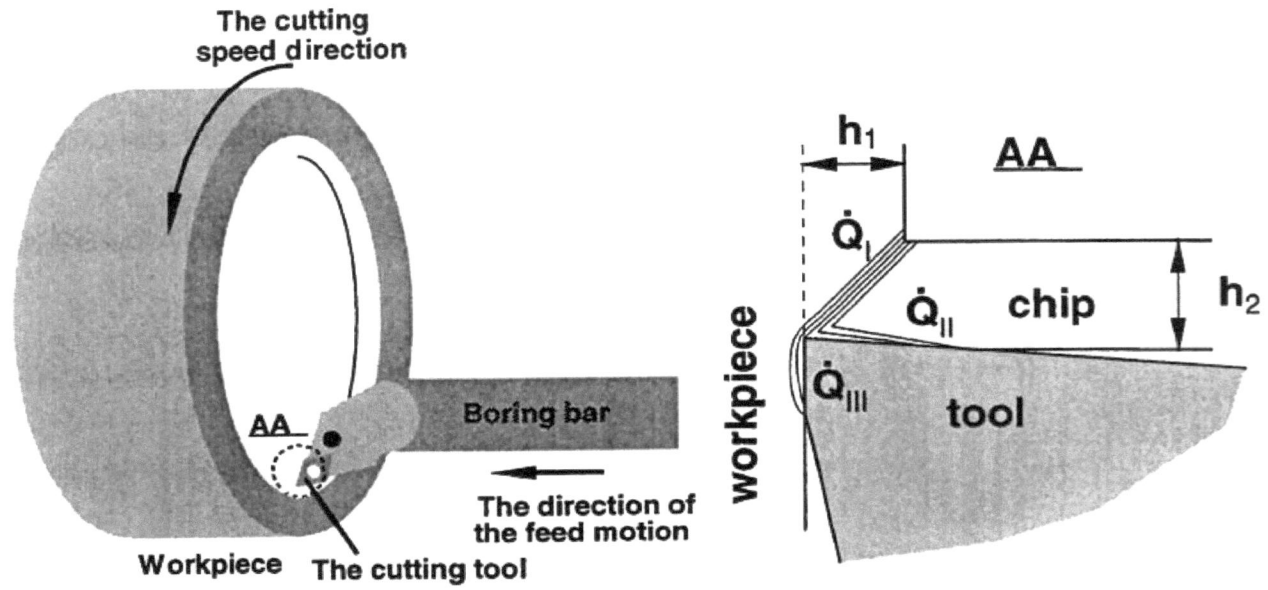

FIG.1.1 GENERAL BORING OPERATION

In internal turning, the metal-cutting process is carried out in pre-drilled holes or holes in cast, etc. The dimensions of the work-piece hole generally determine the length and limit the diameter or cross-sectional size of the boring bar. Usually, a boring bar is long and slender and is thus sensitive to excitation forces introduced by the material deformation process in the turning operation. The boring bar is generally the weakest link in the boring bar–clamping system of the lathe. The boring bar motion may vary with time. This dynamic motion originates from the deformation process of the work material. The motion or vibration of the boring bar influences the result of the machining in general, and the surface finish in particular. Tool life is also likely to be influenced by the resulting vibrations.

The tool vibrations during turning are usually denoted ''self-excited chatter'' or ''tool vibration''. Depending on the driving force of the tool vibration, the vibration is generally divided into one of two categories: regenerative chatter and non-regenerative chatter (primary chatter). Primary chatter may arise from different physical causes, e.g. random excitation of the tool holder's eigenfrequencies due to plastic deformation of work-piece material and/or friction between the tool and the cut material, the tendency of the cutting force excitation to change with the cutting speed, and the dynamic effects of the geometry of the cutting tool on the cutting process, etc. Regenerative chatter is induced by the undulation on the surface of the work piece which is produced during previous successive cuts.

In this investigation, the main objective of the work is to design damped tool for exiting machine tools with low cost.

From the result suggesting which damping material will be suitable for suppression of chatter vibration and temperature produced in tool during boring operations comparing with each other. These results are used in the study of boring tool with and without damper.

CHAPTER-2

2. ELIMINATION OF VIBRATIONS

2.1 Need of Vibration Control

Usually, satellites and other large-scale space structures are light in weight because of mission requirement. For this, they are built up from materials that have low flexural rigidity, very low material damping. Also other forms of damping such as air resistance etc. are absent in their working environment. It may lead to destructive large amplitude vibration and long vibration decay periods and thus result in fatigue, instability and poor operation of the structure. Hence it is needed that there must be some provision in structure to control its vibration.

Vibration control techniques fall into two categories: passive and active. [4] The former requires use of passive components such as vibration dampers and dynamic absorber, which is conventional and well developed. However, the passive control approach suffers from the major drawbacks such as being ineffective at low frequency range and adding significant weight in the system. On the other hand active control approaches provide numerous advantages; e.g. improved low frequency performance, reduction in size and weight and programmable flexibility of design. Thus, in applications mentioned above, active vibration techniques serve as promising alternatives to the conventional passive methods.

2.2 PASSIVE VIBRATION CONTROL SOLUTIONS

Passive vibration control is achieved by incorporating design elements such as damping materials and tuned mass absorbers into a structure to modify its response to initial conditions and/or forced excitations. The phrase passive control arises because no external energy source is associated with its operation. Passive methods have been, and continue to be, the dominant choice for attacking vibration problems because of their simplicity and economy.

2.3 ACTIVE VIBRATION CONTROL SOLUTIONS

Active vibration control is typically achieved by incorporating sensor and actuator pairs in the structural design to modify the response via feedback control. Obviously, once active elements are incorporated into the structure, any type of feedback control may be used, with the caveat that the dynamics we wish to modify are both observable and controllable.

2.4 DAMPING

When designing new machine tools, the design should be such that unwanted vibrations are absent or least reduced. During the design of the machine tool should make it a point to consider all the possibilities of vibration elimination. The overall stiffness of the machine tools is very important. The stiffness and damping capacity are two predominant features for determining dynamic characteristics of the machine tools and components. Dimensional accuracy is dependent on static stiffness and should be tackled at the damping should be introduced from the conceptual design stage. To reduce intensity of vibration, it is also necessary that damping should be introduced from the conceptual stage of design stage.

In machine tool, there are mainly there sources of damping. First, damping occurs due to internal friction of the vibrating structure, i.e. structure damping secondly, energy is dissipated due to the rubbing between surfaces of two elements at the junction during vibration and damping occurs due to the presence of surrounding medium air. Of course, the magnitude of last one is very small comparison to the magnitude of first two.

2.5 PASSIVE DAMPING

For an example Vibration absorber which comprise of a parallel spring damper combination which couples the main structure to an auxiliary mass. Removing the coupling spring the absorber system results in the well-known 'Lanchester damper' and if the damping element is removed 'impact damper' is obtained. This group of dampers is categorized as passive dampers .The basic principle of vibration absorber depends on the principle of transferring vibrational energy to an auxiliary system.

2.6 ACTIVE DAMPING

Active vibration control in boring operations clearly is a possible solution to reduce the vibrations present in this kind of machining. Embedding the actuator and accelerometer into the boring bar enables the design to be applicable to a general lathe as long as the mounting arrangement is fairly similar. Embedding the electronic devices also protects them from the harsh environment in a lathe. The metal chips from the cutting process and the cutting fluids would otherwise constitute big problems to the actuator and accelerometer.

CHAPTER-3

3. TOOL USED FOR ANALYSIS

3.1 INTRODUCTION TO PRO-E

Pro/ENGINEER is a powerful program used to create complex designs with a great precision. The design intent of any 3-D model or an assembly is defined by its specification and its use. The powerful tools of Pro/ENGINEER are to capture the design intent of any complex model by incorporating intelligence into the design.

To make the designing process simple and quick, this software package has divided the steps of designing into different modules. This means each step of designing is completed in a different module. For example, generally a design process consists of the following steps:

1. Sketching using the basic sketch entities.
2. Converting the sketch features and paths.
3. Assembling different parts and analyzing them.
4. Documentation of the parts and the assembly in terms of drawing views.
5. Manufacturing the final part and assembling.

Feature Based Nature of Pro/Engineer

Pro/ENGINEER is a feature-based solid modeling tool. A feature is defined as the smallest building block and any solid model created in Pro/ENGINEER is an integration of number of these building blocks. Each feature can be edited individually to bring in any change in the solid model. This feature provides greater flexibility. Pro/E allows make modifications by just modifying some values in the same part.

Bidirectional Associative Nature of Pro/Engineer

There is bidirectional associativity between all modes of Pro/ENGINEER. The bidirectional associative nature is defined as its ability to ensure that if any modification in particular model in one mode, the modification is reflected in the same model in other modes also. For example any change in the Assembly Mode, after regeneration, the change will be highlighted in the Part Mode also. This feature correlates the 2-D drawing views in the Drawing Mode and the solid model created in the Part Mode of Pro/ENGINEER and vise versa. The bidirectional associativity means that if modification is made to any one application, it changes the output of all other modes related to the model.

Parametric Nature of Pro/Engineer

Pro/ENGINEER is parametric in nature. The parametric nature of Pro/ENGINEER means that the features of a part become interrelated if they are drawn by taking the reference of each. It allows redefining the dimensions or the attributes of a feature at any time. This will develop the relationship among them. This relationship is known as parent child relationship. Changing the child feature will alter the reference design as per requirement

3.2 ANSYS ANALYSIS

It is a Finite Element Analysis software enables to perform the tasks:

1. Build models or Transfer CAD models of structures products components or system.
2. Apply operating loads or design performance conditions.
3. Study the physical responses, such as stress levels, temperature distribution, or the impact of Electromagnetic fields.
4. Optimize a design in the development process to reduce production cost.
5. To prototype testing in environments where it otherwise could be undesirable or impossible.

The ANSYS program a comprehensive Graphical User Interface (GUI) that gives user an easy interactive access to program functions, commands, and documentation and reference material. A utility menu system helps user to navigate through the ANSYS program.

Creating the Model

The model is drawn in 1D, 2D or 3D space in the appropriate units (m, mm, in, etc). The model may be created in the pre-processor, or it can be imported from another CAD drafting package (like **PRO-E)** via a neutral file format. If a model is drawn in mm for example and the material properties are defined in SI units, then the results will be out of scale by factors of 10^6. The same units should be applied in all directions, otherwise results will be difficult to interpret, or in extreme cases the results will not show up mistakes made during the loading and restraining of the model.

Nodes and Elements

Node is a coordinate location in space where degrees of freedom and actions of the physical system exist.Element is mathematical, matrix representation (called stiffness or coefficient matrix) of the interaction among the degrees of freedom of a set of nodes. Elements may be line, area, or solid and two or three-dimensional.

The Finite Element Analysis model consists of a number of simply shaped elements, connected to nodes, Information is passed from element to element only at common nodes.

Creating a Mesh

Mesh generation is the process of dividing the analysis continuum into a number of discrete parts or finite elements. Finer the mesh, better the result but which longer the analysis time. In the manually created mesh, you will notice that the elements are smaller at the joint. This is known as mesh refinement, and it enables the stresses to be captured at the geometric discontinuity (the junction). Manual meshing is a long and tedious process for models with any degree of geometric complication, but with useful tools emerging in pre-processors, the task is becoming easier.

Automatic mesh generators are very useful and popular. A mesh engine creates the mesh automatically; the only requirement is to define the mesh density along the model's edges. Automatic meshing has limitations as regards mesh quality and solution accuracy.

Assigning Properties

Material properties (Young's modulus, Poisson's ratio, the density, and if applicable, coefficients of expansion, friction, thermal conductivity, damping effect, specific heat etc.) will have to be defined. If 2D elements are being used, the thickness property is required. One Dimensional beam elements require area, Ixx, Iyy, J a direction cosine property, which defines the direction of the beam axis in 3D space. Shell elements, which are 2½D in nature (2D elements in 3D space), require orientation neutral surface offset parameters to be defined.

Apply Loads

Some type of load is usually applied to the analysis model. The loading may be in the form of a point load, a pressure or a displacement in a stress (displacement) analysis, a temperature or a heat flux in a thermal analysis and a fluid pressure or velocity in a fluid analysis. The loads may be applied to a point, an edge, a surface or even a complete body. The loads should be in the same units as the model geometry and material properties specified. In the cases of modal (vibration) and buckling analyses, a load does not have to be specified for the analysis to run.

Applying Boundary Conditions

If you apply a load to the model, then in order to stop it accelerating infinitely through the computer's virtual ether (mathematically known as a zero pivot), at least one constraint or boundary condition must be applied. Structural boundary conditions are usually in the form of zero displacements, thermal Boundary Condition are usually specified temperatures, fluid Boundary Condition are usually specified pressures. A boundary condition may be specified to act in all directions (x, y, z), or in certain directions only. They can be placed on nodes, key points, and areas or on lines. At least one Boundary Condition has to be applied to every model, even modal and buckling analysis with no loads applied.

Solution

The Finite Element Analysis solver can be logically divided into three main parts, the pre-solver, the mathematical-engine and the post-solver. The pre-solver reads in the model created by the pre-processor and formulates the mathematical representation of the model. If the model is correct the solver proceeds to form the element-stiffness matrix for the problem and calls the mathematical-engine, which calculates the result (displacement, temperatures, pressures, etc.). The results are returned to the solver and the post-solver is used to calculate strains, stresses, heat fluxes, velocities, etc.) for each node within the component or continuum. All these results are sent to a results file, which may be read by the post-processor.

Post-Processor

Here the results of the analysis are read and interpreted. They can be presented in the form of a table, a contour plot, deformed shape of the component or the mode shapes and natural frequencies if frequency analysis is involved. Other results are available for fluids, thermal and electrical analysis types. Most post-processors provide an animation service, which produces an animation and brings your model to life.

3.2.1 STRUCTURAL ANALYSIS

This type of analysis is the most common application of Finite Element Analysis and is used primarily for mechanical and civil engineering applications. Structural analysis is possible in the following areas.

Static Analysis - Static analysis determines the displacements, stresses, strains, and forces in structures or components caused by loads that do not induce significant inertia and damping effects.

Modal Analysis - used to calculate natural frequencies and mode shapes of a structure.

Harmonic Analysis - used to determine the harmonic response of a structure to time-varying loads.

Transient Dynamic Analysis - used to determine the response of a structure to random time-varying loads.

Spectrum Analysis - used to calculate stresses and strains of a structure due to a response spectrum or a random vibration input.

Buckling Analysis - used to calculate buckling loads and determines the buckling load shape.

Heat Transfer Analysis – To compute the temperature distribution and heat flow within a structure under a steady state and transient conditions.

Field Problems – Analysis of field problem in acoustics and fluid mechanics.

Coupling Effects – Solution techniques for inter facing multiple field effects such as, displacement forces, temperature, heat flows, electrical voltage and current, magnetic field intensity and flux, fluid pressure and velocity.

3.2.2 THERMAL ANALYSIS

A thermal analysis calculates the temperature distribution and related thermal quantities in a system or component. Typical thermal quantities of interest are:

- The temperature distributions.
- The amount of heat lost or gained.
- Thermal gradients.
- Thermal fluxes.

Types of Thermal Analysis

A **steady-state** thermal analysis determines the temperature distribution and other thermal quantities under steady-state loading conditions. A steady-state loading condition is a situation where heat storage effects varying over a period of time can be ignored.

A **transient** thermal analysis determines the temperature distribution and other thermal quantities under conditions that vary over a period of time.

Tasks in a Thermal Analysis

The procedure for doing a thermal analysis involves three main tasks:

- Build the model.
- Apply loads and obtain the solution.
- Review the results.

CHAPTER-4

4. TOOL GEOMETRY AND MATERIAL PROPERTIES

4.1 DESIGN CONSIDERATION OF A TOOL

- When designing new machine tool, the design should be such that unwanted vibrations are absent or least reduced.

- During the design of the machine tool ,the possibilities of elimination of vibration should be considered.

- The stiffness and damping capacity are two predominant features for determining dynamic characteristics of the machine tools.

4.2 TOOL AND INSERT GEOMETRY

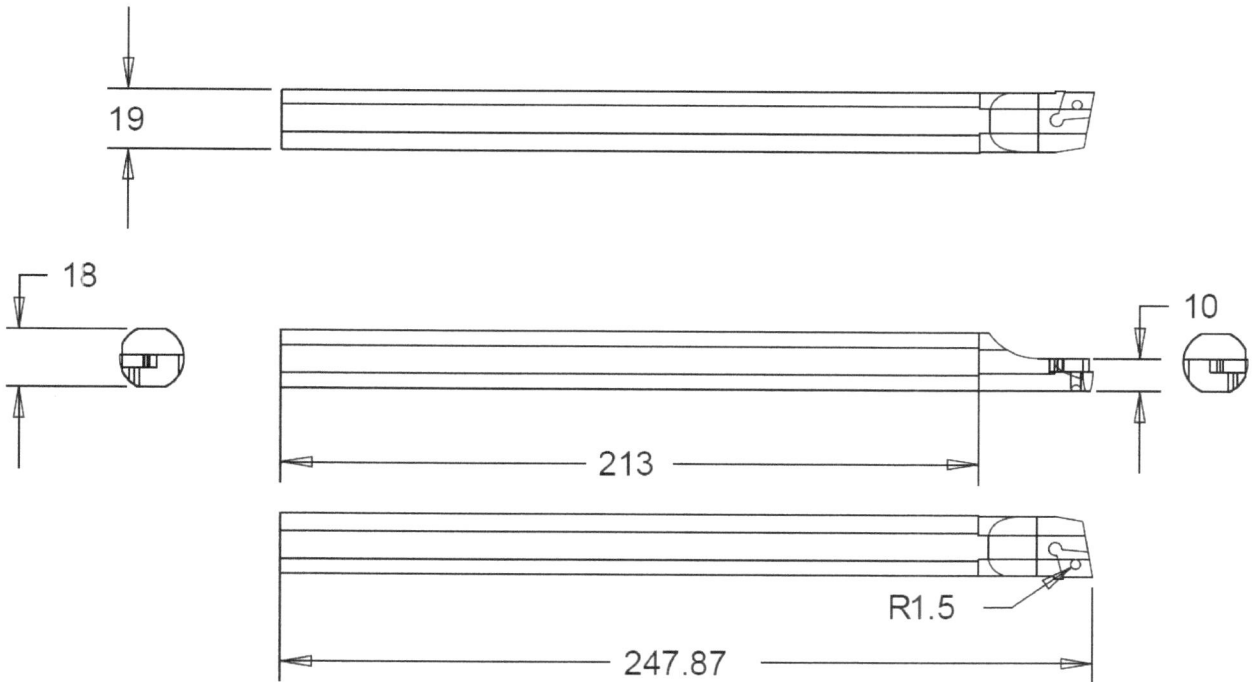

ALL DIMENSIONS ARE IN mm   FIG 4.2.1 GEOMETRY OF TOOL

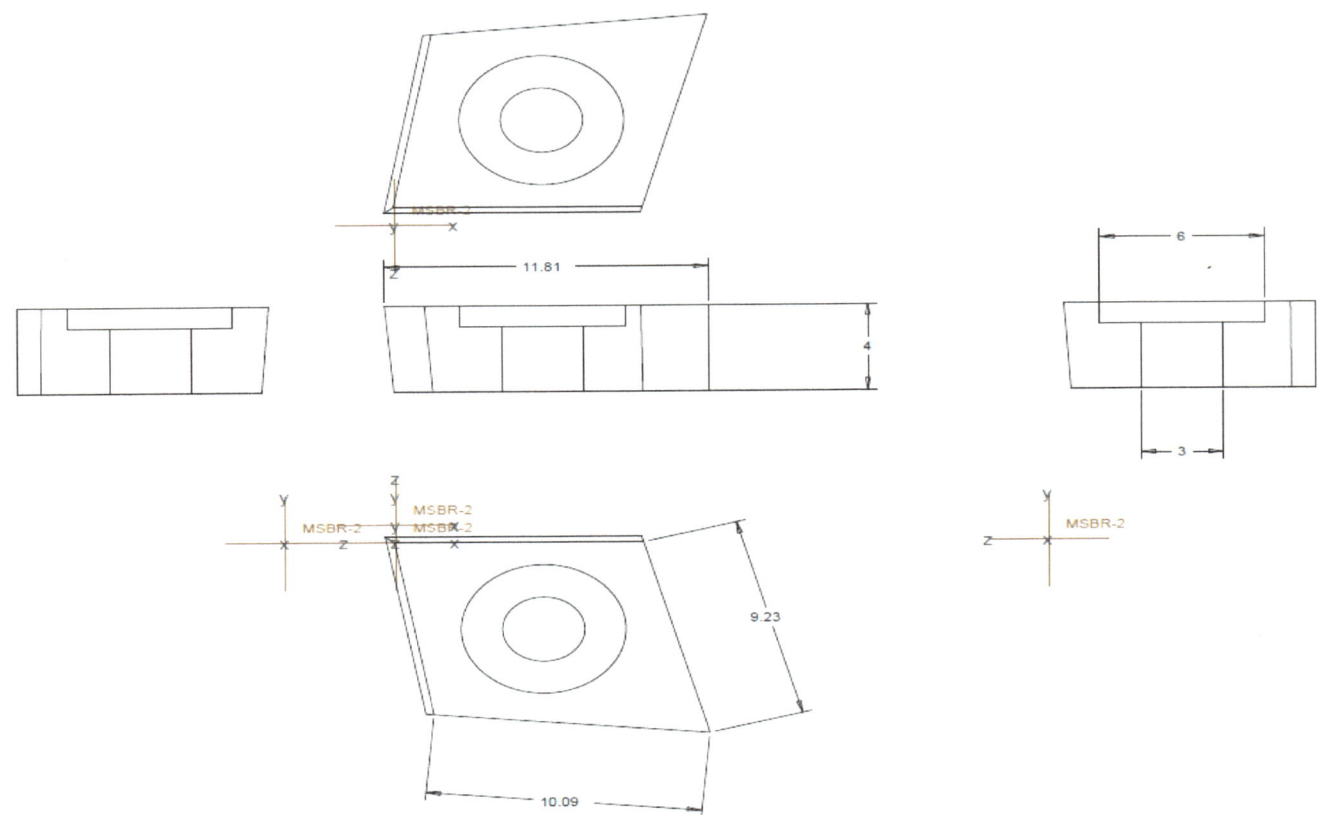

ALL DIMENSIONS ARE IN mm

FIG 4.2.2 GEOMETRY OF INSERT

4.3. SOLID MODEL OF DAMPER

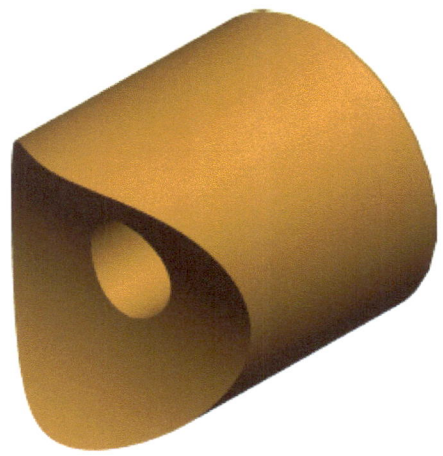

FIG 4.3.1 SOLID OF DAMPER

4.4 SELECTION OF MATERIAL

The best material is one which will serve the desired purpose at minimum cost. The factors which should be considered while selecting the material for machine tool and damping material are as follows:

i. Availability
ii. Cost
iii. Mechanical properties
iv. Manufacturing consideration.

4.5 TOOL PROPERTY

Material - EN31

Density - 7.84×10^{-6} kg/mm^3

Thermal conductivity - 46.6W/mK

Young's modulus - 2.84×10^5 N/mm^2

Poison's ratio - 0.3

Material	C%	Mn%	P_{max}%	S_{max}%	Si_{max}%	Cr%
AISI 52100	0.98-1.1	0.25-0.45	0.025	0.025	0.15-0.3	1.3-1.6

4.6 INSERT PROPERTY

Material - cemented carbide, 6% co

Density	-	14.95×10^{-6} kg/mm^3
Thermal conductivity	-	100 W/mK
Young's modulus	-	6.25×10^5 N/mm^2
Poisson's ratio	-	0.22

4.7 DAMPING MATERIAL PROPERTIES

4.7.1 BRASS:

Brass has a muted yellow color, somewhat similar to gold.Brass is a substitutional alloy. It is used for decoration for its bright gold-like appearance; for applications where low friction is required such as locks, gears, bearings, doorknobs, ammunition, and valves; for plumbing and electrical applications; and extensively in musical instruments such as horns and bells for its acoustic properties

Density	-	8.49×10^{-6}kg/mm^2
Thermal conductivity	-	115 W/mK
Specific heat	-	380 J/kgC

4.7.2 COPPER:

Copper is a ductile metel with very high thermal and electrical conductivity. Pure copper is rather soft and malleable and a freshly-exposed surface has a pinkish or peachy color.Gold,Caesium and copper are the only metallic elements with a natural color other than gray or white. It is used as a thermal conductor, an electrical conductor, a building material, and a constituent of various metal alloys.

Density	-	8.92 x10^{-6} kg/mm²
Young's modulus	-	1.30 x10^5 N/mm²
Thermal conductivity	-	400100 W/mK
Poisson's ratio	-	0.34

4.7.3 CASTIRON

Density	-	7.92 x10^{-6} kg/mm²
Young's modulus	-	1.1 x10^5 N/mm²
Thermal conductivity	-	52 W/mC
Poisson's ratio	-	0.28.

CHAPTER 5

EXPERIMENTAL ANALYSIS

5.1 EXPERIMENTAL METHOD

The boring operation is carried out in lathe.

- The tool is fixed in the lathe 100 mm of distance from the insert tip.

- Boring operation is carried out with and without damper .

- Vibration produced in tool during operation is measured with vibrometer.

- The cutting force of tool during operation is measured by DYNOWARE..

- Temperature produced in tool during operation is measured by

  DA100 (DATA ACQUISITION UNIT).

FIG 5.1 EXPERIMENTAL SETUP

Fig.5.2 Boring Tools & Dampers

Fig.5.3 Without Damping Tool

Fig.5.4. Boring Tool with Insert Damper

5.2 TEMPERATURE GRAPHS

The graphs were obtained from YOKOGAWA software show the boring tool tip temperature which was sensed by chromel-alumel thermo couple during operation. The experiments were conducted without damping and with damping. In this graphs, temperature was along Y-axis and time was along X-axis.

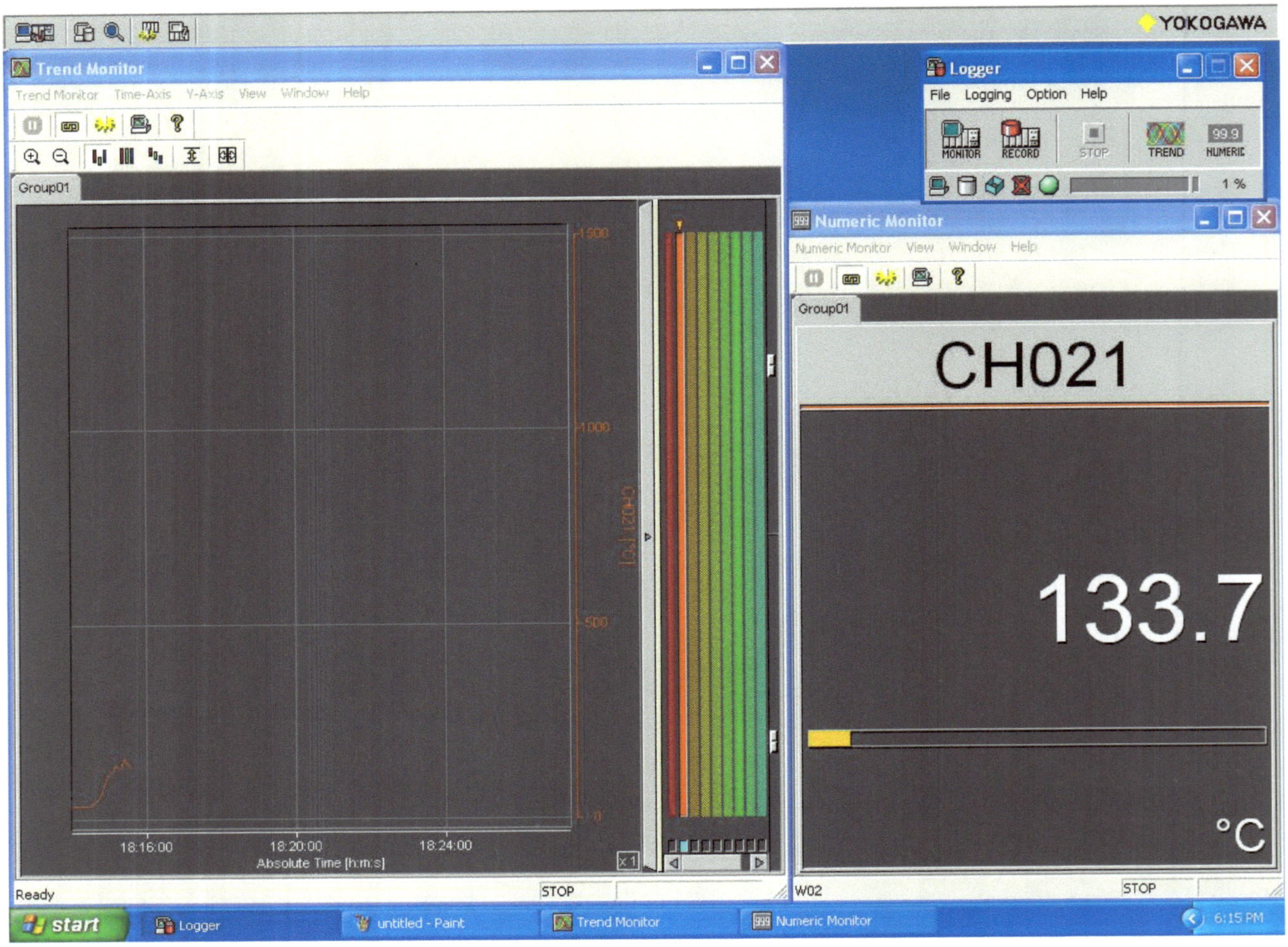

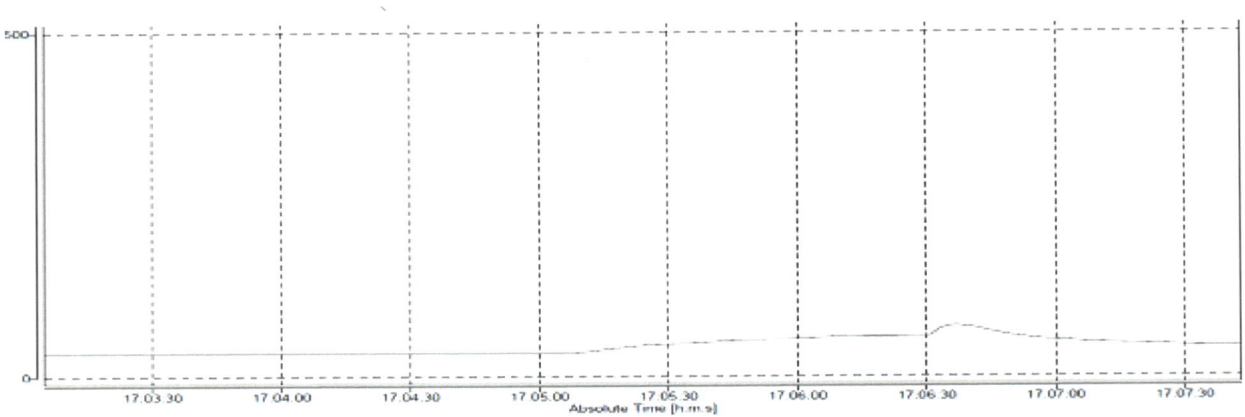

Fig 5.5 SOFTWARE PLATFORM FOR DRAWING GRAPH

GRAPH 5.1- WITHOUT DAMPING FOR 0.01mm AND 500rpm

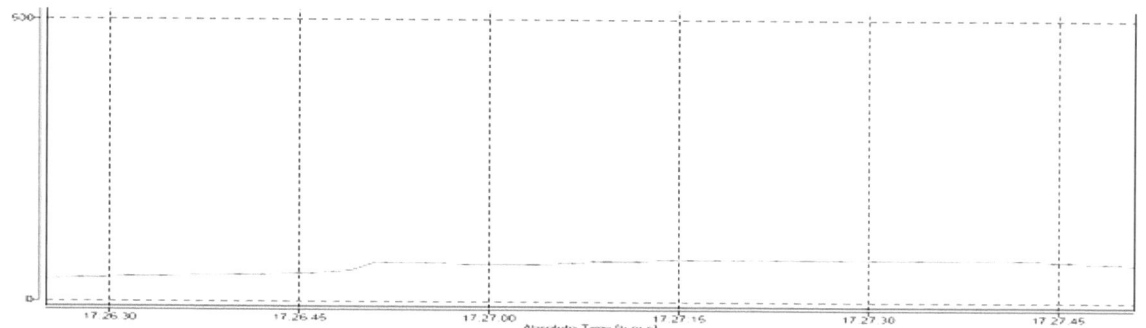

GRAPH 5.2-WITHOUT DAMPING FOR 0.01mm AND 600rpm

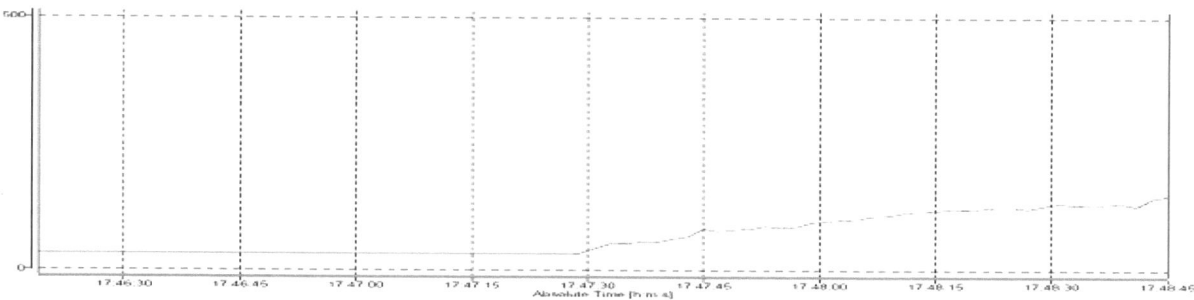

GRAPH 5.3 WITHOUT DAMPING FOR 0.02mm AND 500rpm

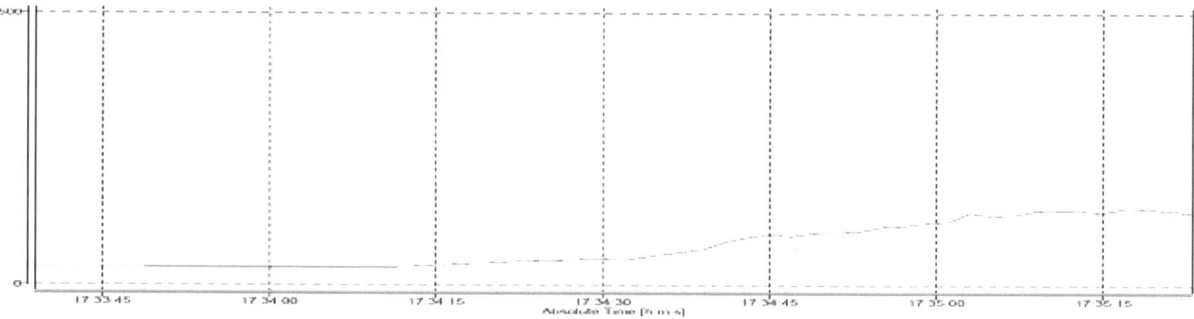

GRAPH 5.4 WITHOUT DAMPING FOR 0.02mm AND 600rpm

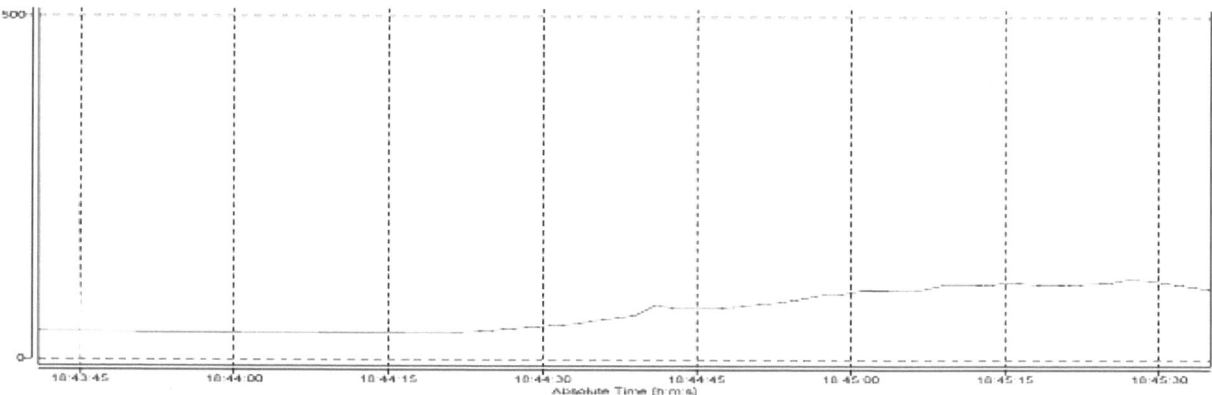

GRAPH 5.5 WITH DAMPER- BRASS FOR 0.01mm,500rpm

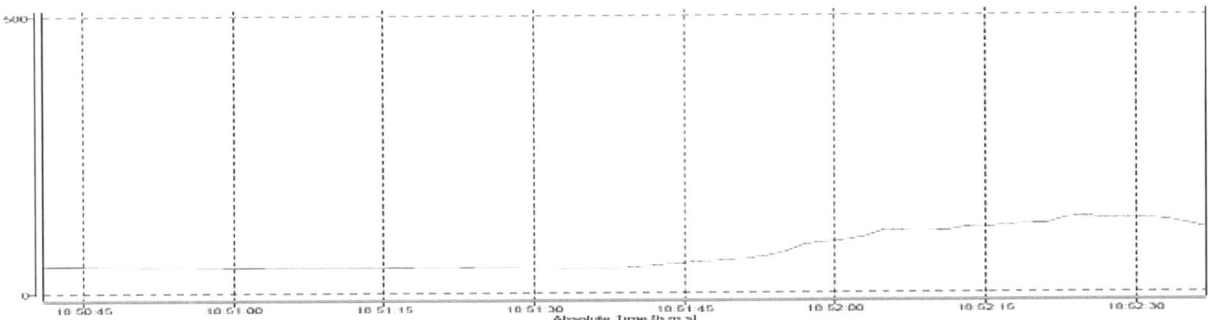

GRAPH 5.6 WITH DAMPER-BRASS FOR 0.01mm and 600rpm

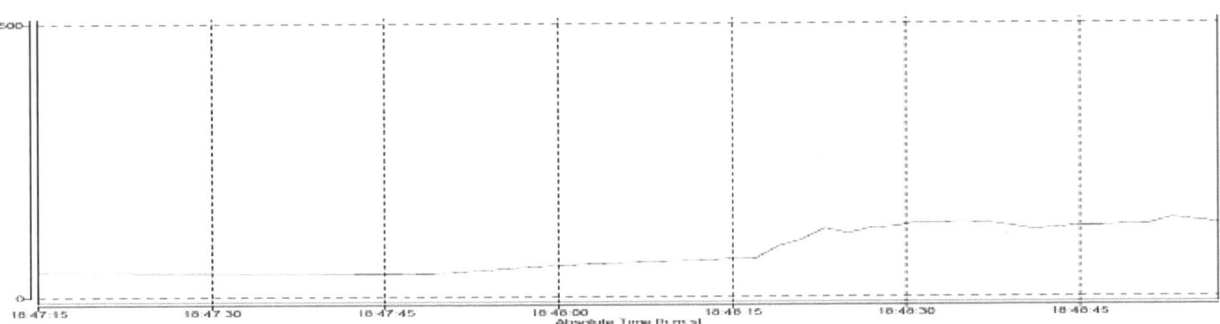

GRAPH 5. 7 WITH DAMPER-BRASS FOR 0.02mm and 500rpm

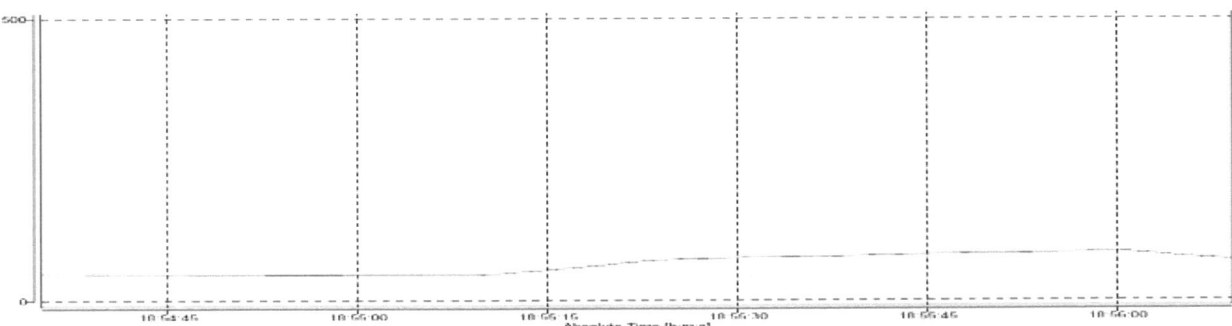

GRAPH 5.8 WITH DAMPER-BRASS FOR 0.02mm and 600rpm

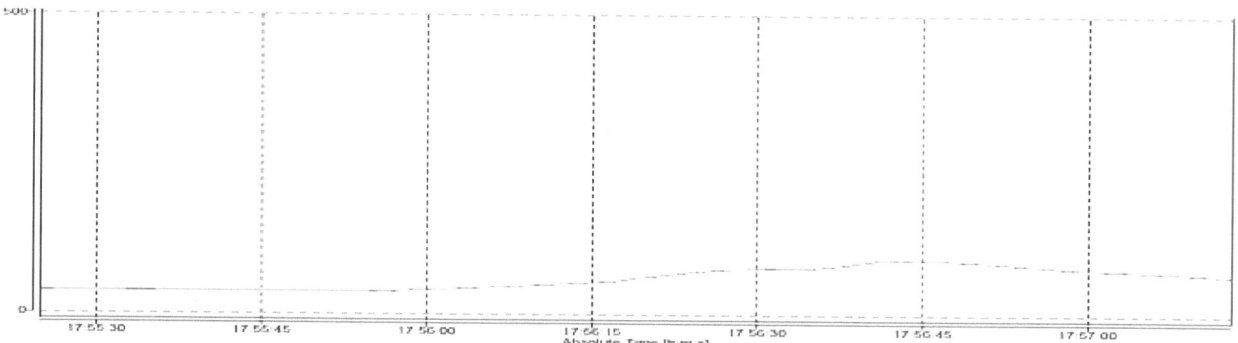

GRAPH 5.9 WITH DAMPER-EN31 FOR 0.01mm and 500rpm

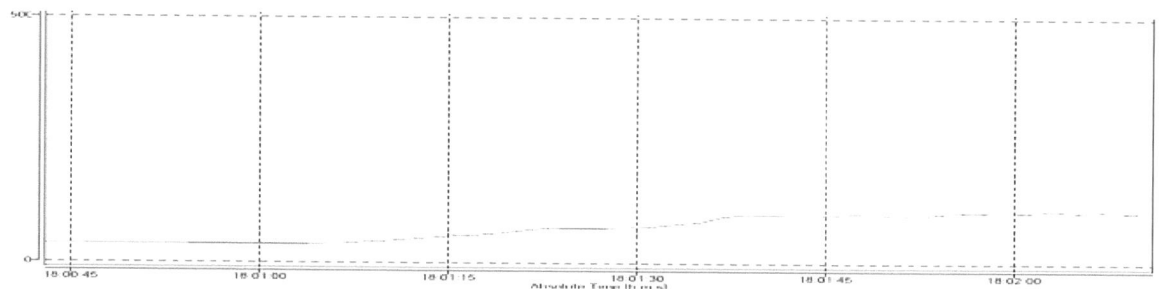

GRAPH 5.10- WITH DAMPER-EN31 FOR 0.01mm and 600rpm

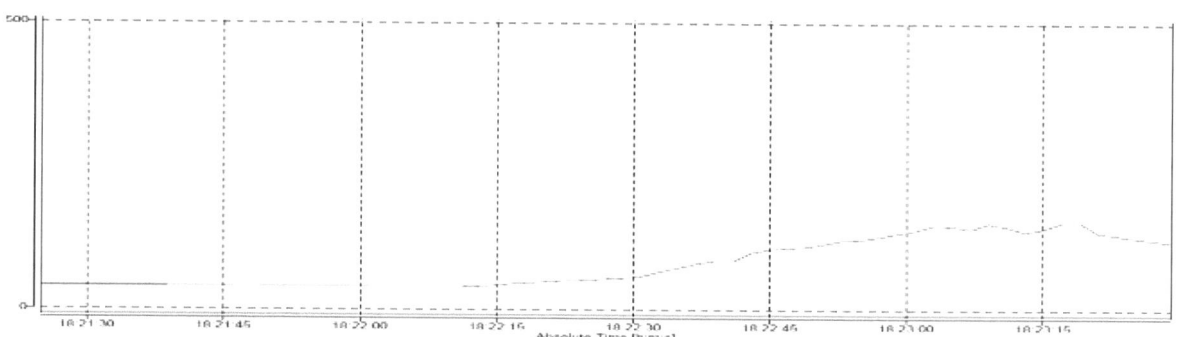

GRAPH 5.11- WITH DAMPER-EN31 FOR 0.02mm and 500rpm

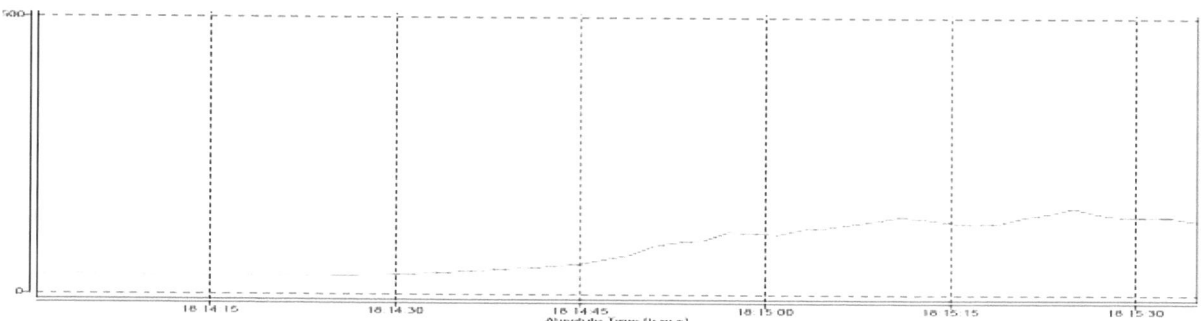

GRAPH 5.12- WITH DAMPER-EN31 FOR 0.02mm and 600rpm

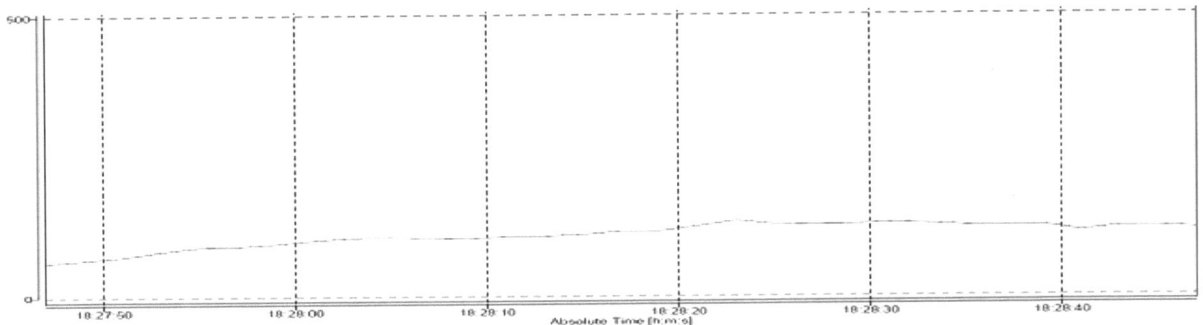

GRAPH 5.13 WITH DAMPER-CASTIRON FOR 0.01mm and 500rpm

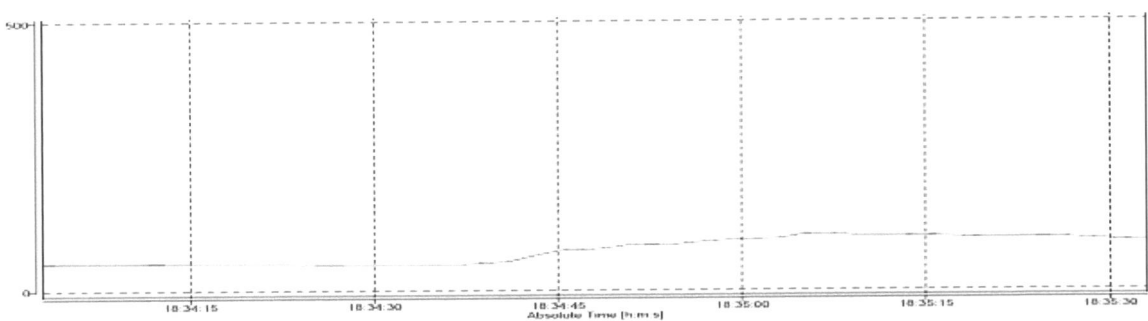

G RAPH 5.14- WITH DAMPER-CASTIRON FOR 0.01mm and 600rpm

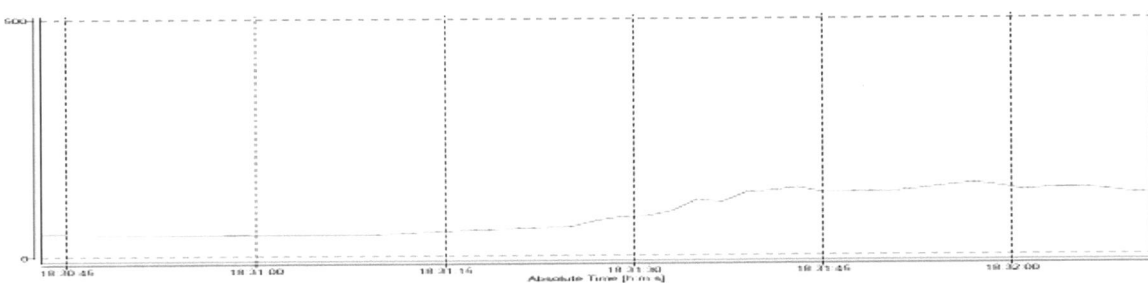

GRAPH 5.15- WITH DAMPER-CASTIRON FOR0.02mm and 500rpm

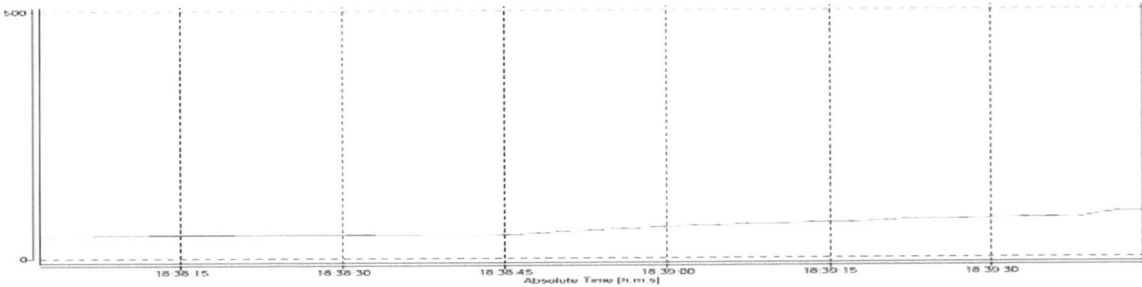

GRAPH 5.16- WITH DAMPER-CASTIRON FOR 0.02mm and 600rpm

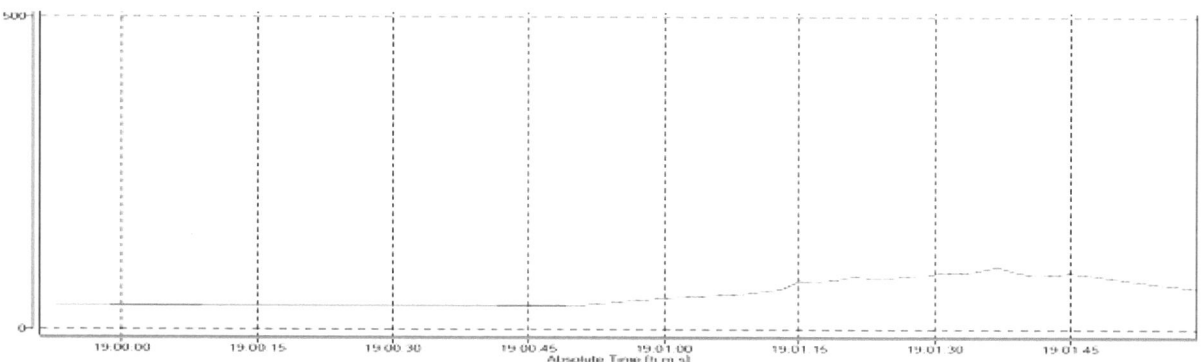

GRAPH 5.17- WITH DAMPER-COPPER FOR 0.01mm and 500rpm

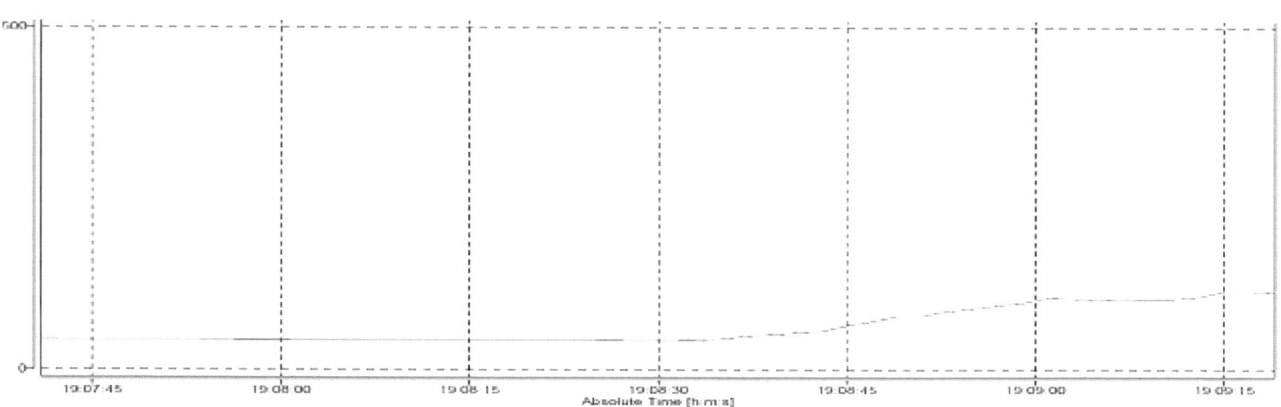

GRAPH 5.18- WITH DAMPER-COPPER FOR 0.01mm and 600rpm

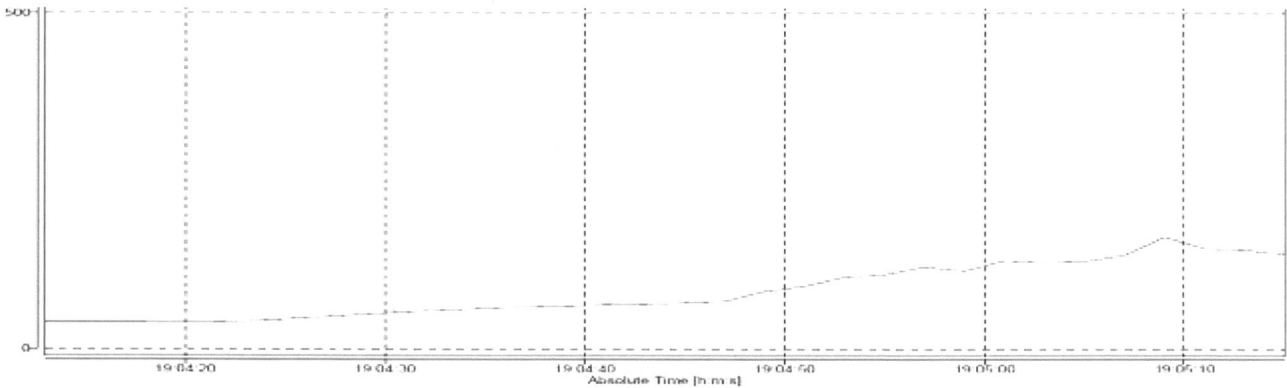

GRAPH 5.19- WITH DAMPER-COPPER FOR 0.02mm and 500rpm

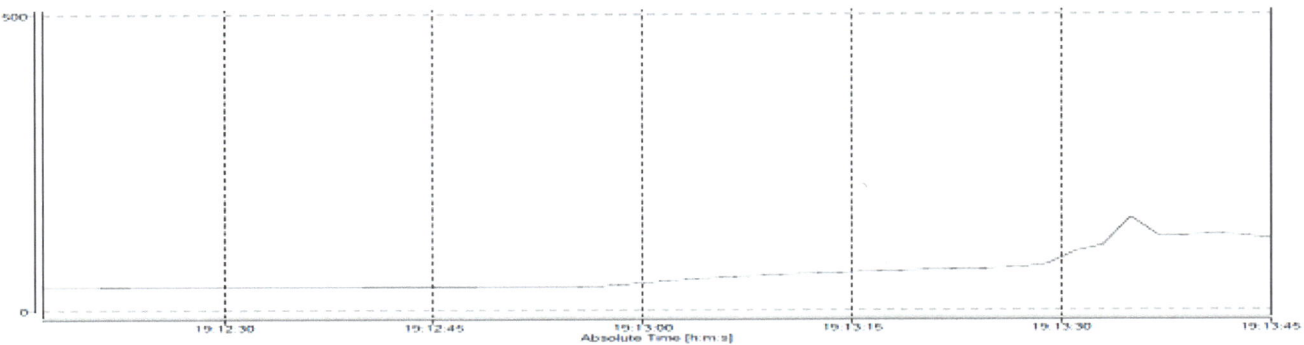

GRAPH 5.20- WITH DAMPER-COPPER FOR0.02mm and 600rpm

5.3 CUTTING FORCE GRAPHS

Cutting forces are Measured and are shown as graph by dynoware data acquisition unit.

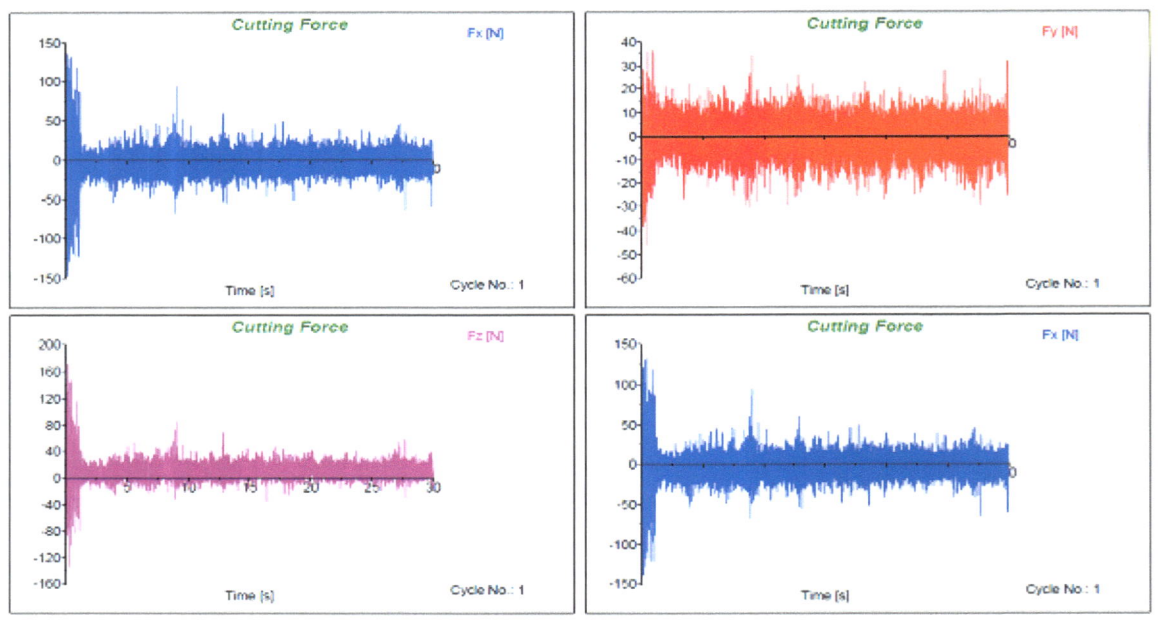

GRAPH 5.21- WITHOUT DAMPING FOR 0.01mm DEPTH, 500rpm

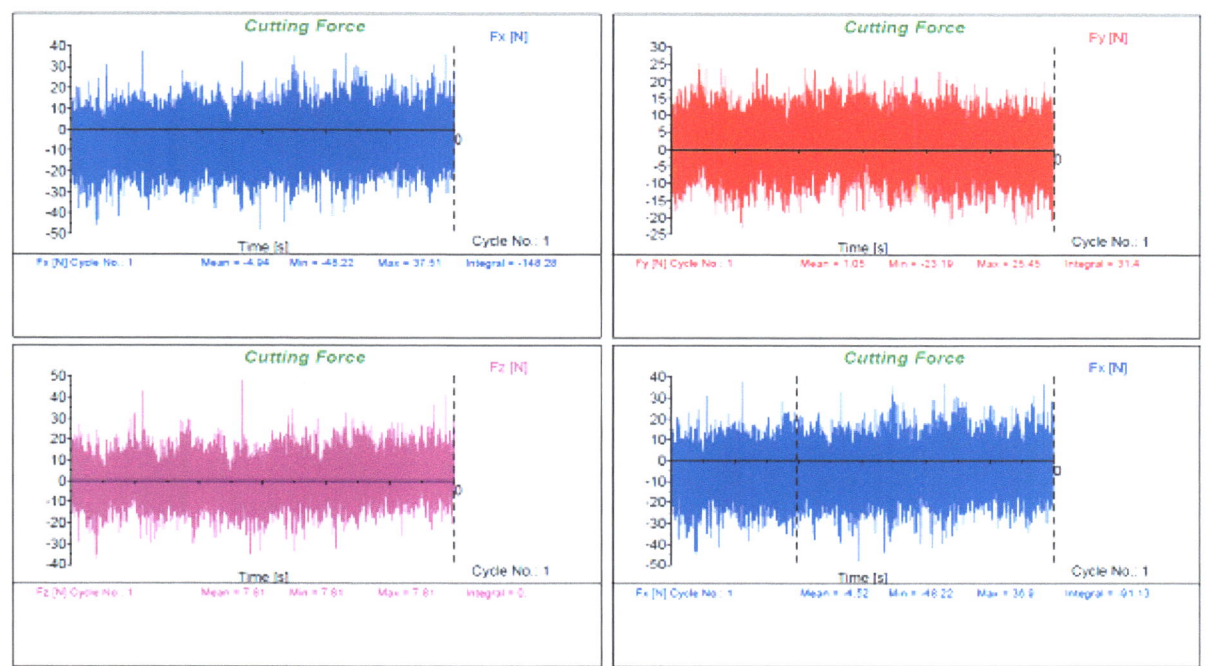

GRAPH 5.22- WITHOUT DAMPING FOR 0.01mm DEPTH, 600rpm

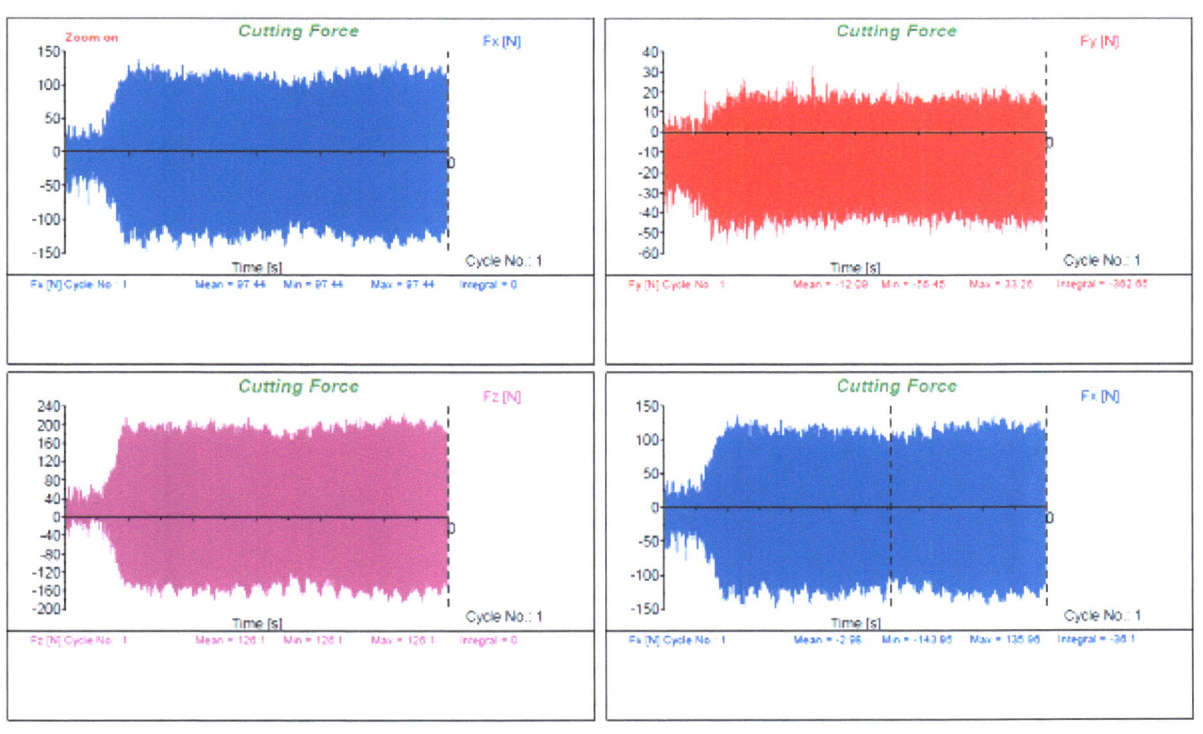

GRAPH 5.23- WITHOUT DAMPING FOR 0.02mm DEPTH, 500rpm

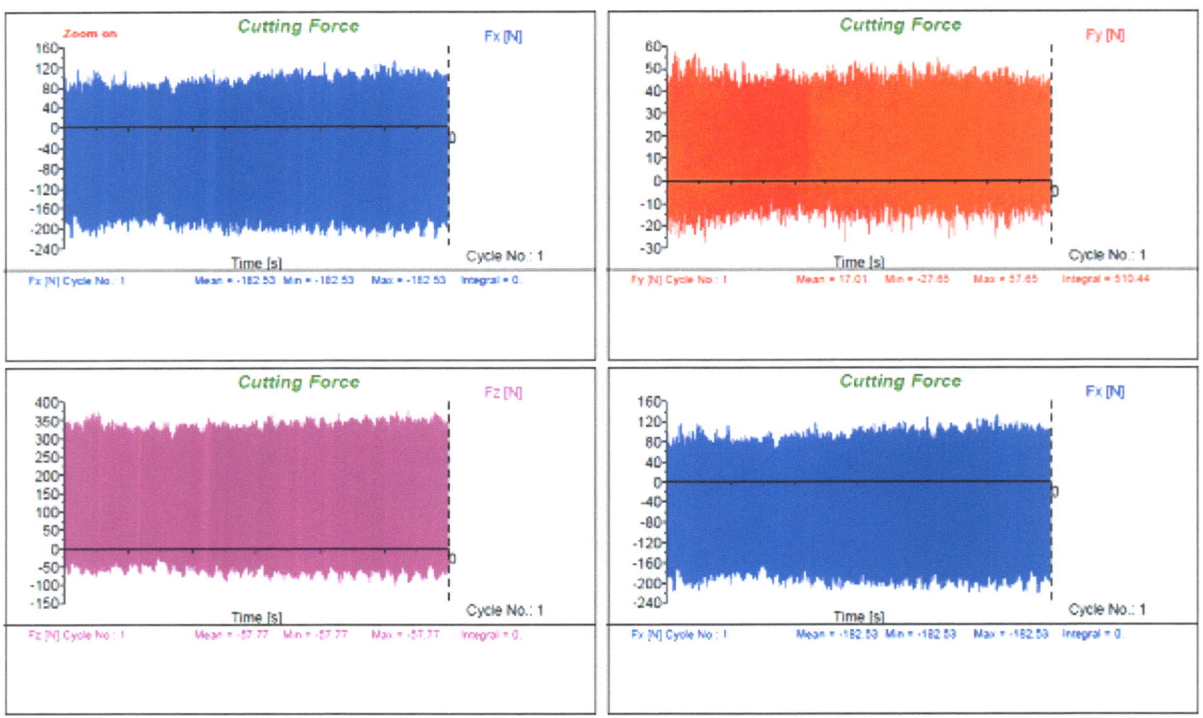

GRAPH 5.24- WITHOUT DAMPING FOR 0.02mm DEPTH, 600rpm

CUTTING FORCE GRAPH FOR THE BORING TOOL WITH DAMPING
MATERIAL OF BRASS

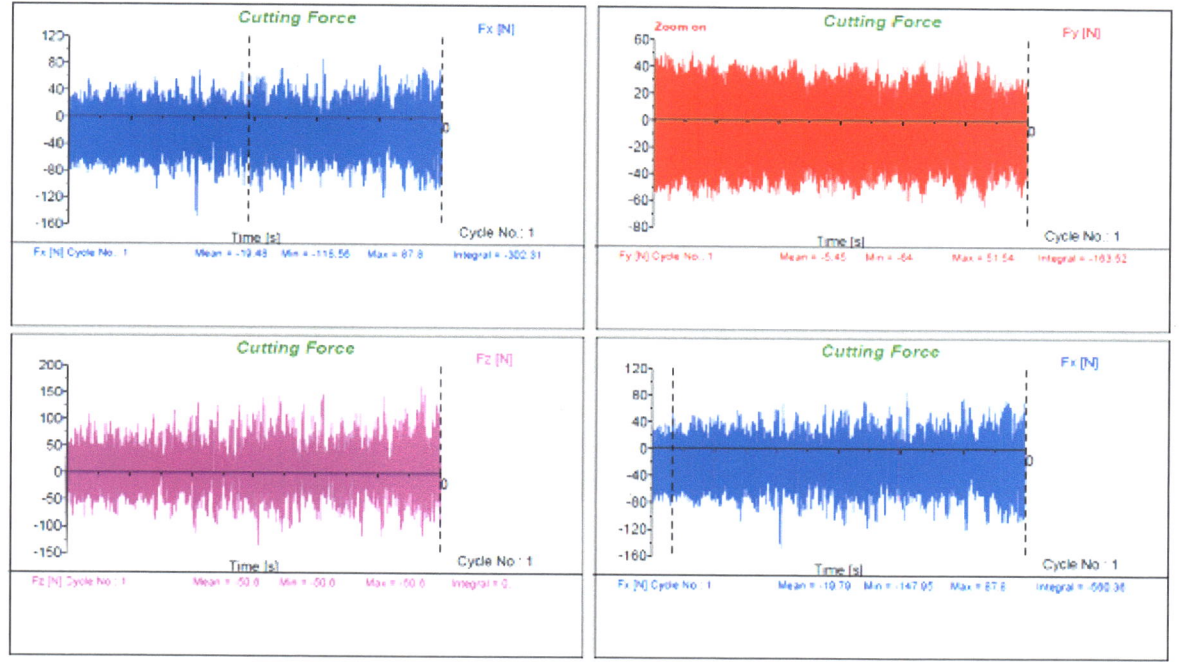

GRAPH 5.25 WITH DAMPER- BRASS FOR 0.01mm DEPTH, 500rpm

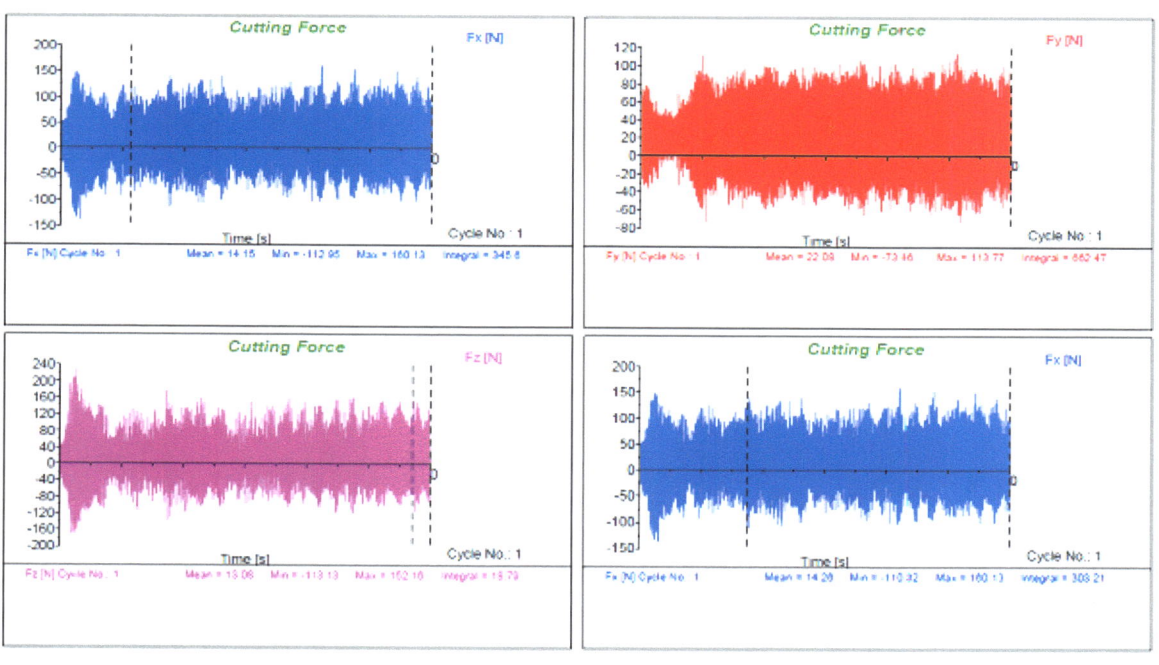

GRAPH 5.26 WITH DAMPER- BRASS 0.01mm DEPTH, 600rpm

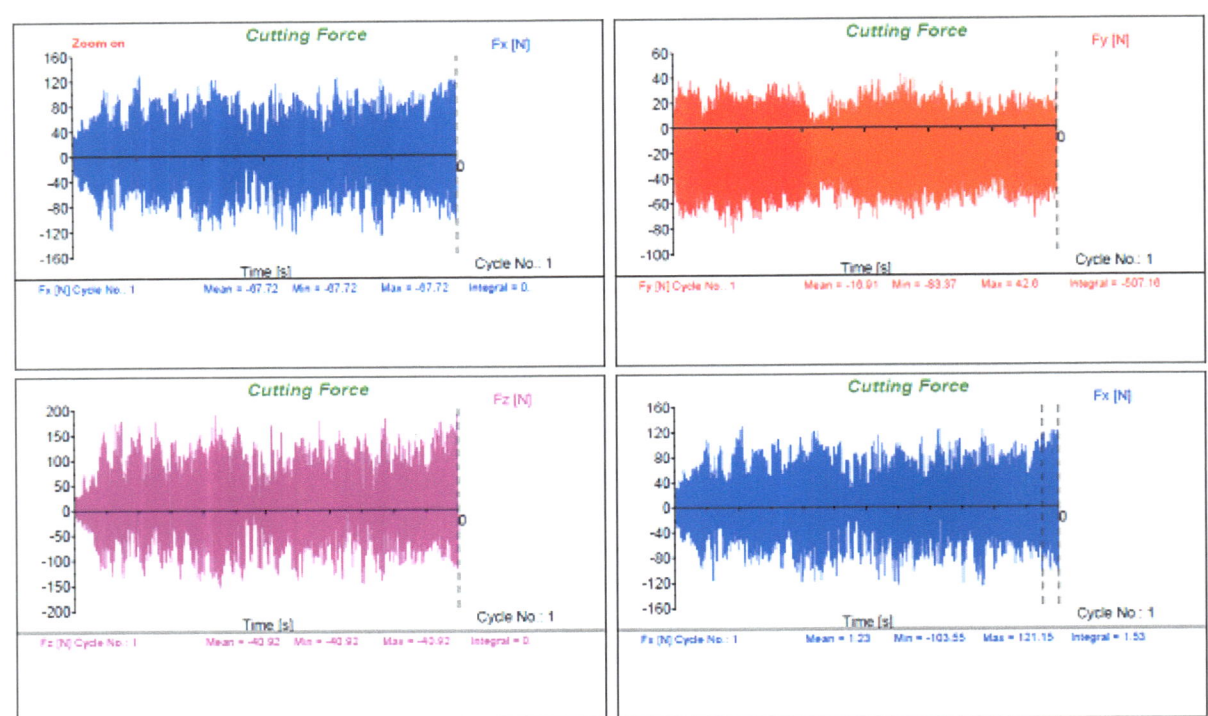

GRAPH 5.27 WITH DAMPER- BRASS0.02mm DEPTH, 500rpm

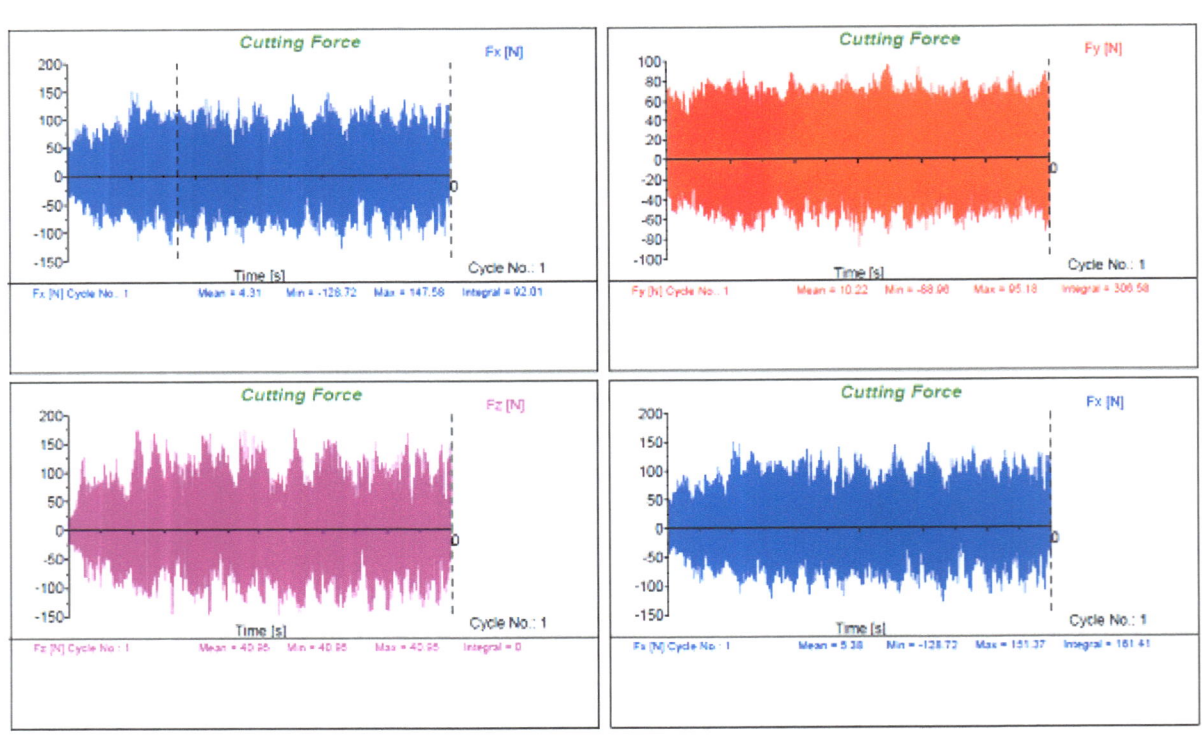

GRAPH 5.28 WITH DAMPER- BRASS 0.02mm DEPTH, 600rpm

CUTTING FORCE GRAPH FOR THE BORING TOOL WITH DAMPING MATERIAL OF CASTIRON

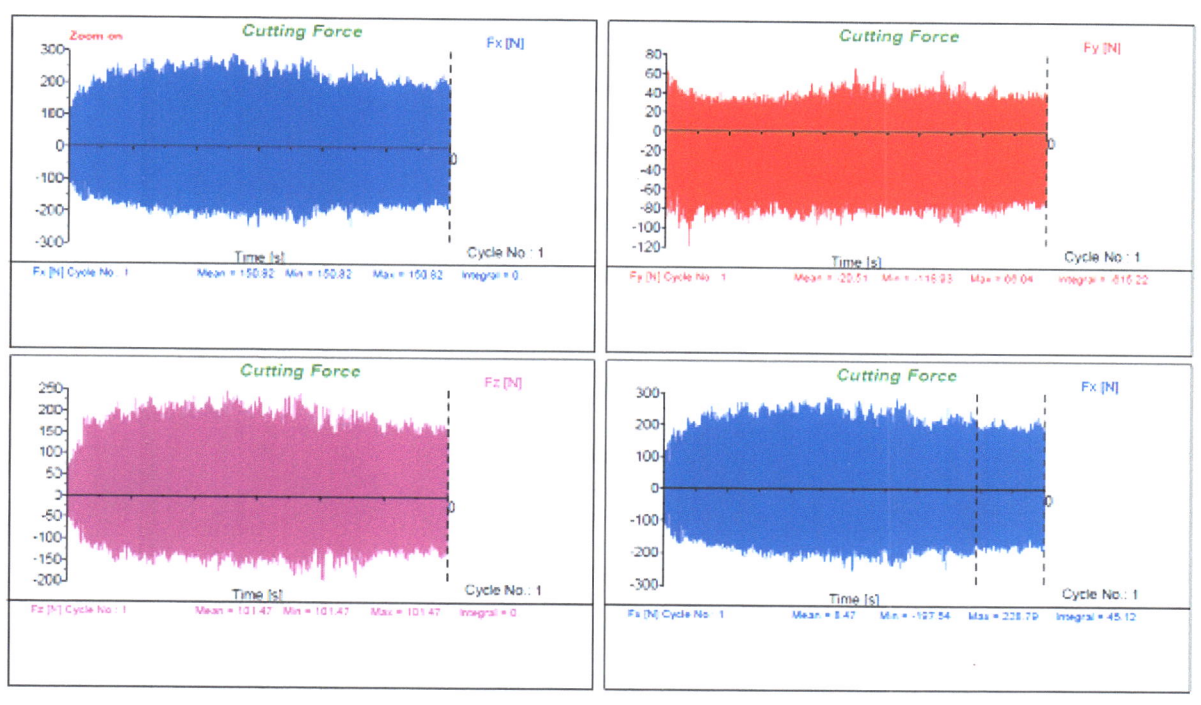

GRAPH 5.29- WITH DAMPER-CASTIRON FOR 0.01mm DEPTH, 500rpm

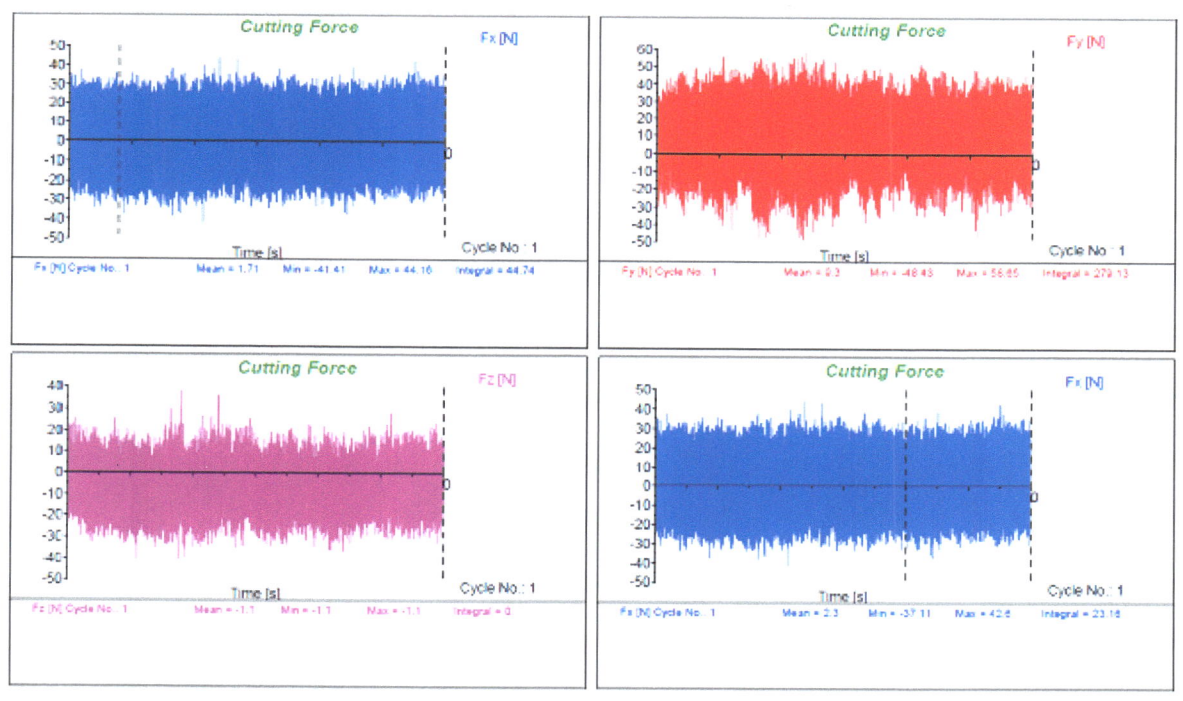

GRAPH 5.30- WITH DAMPER-CASTIRON FOR 0.01mm DEPTH, 600rpm

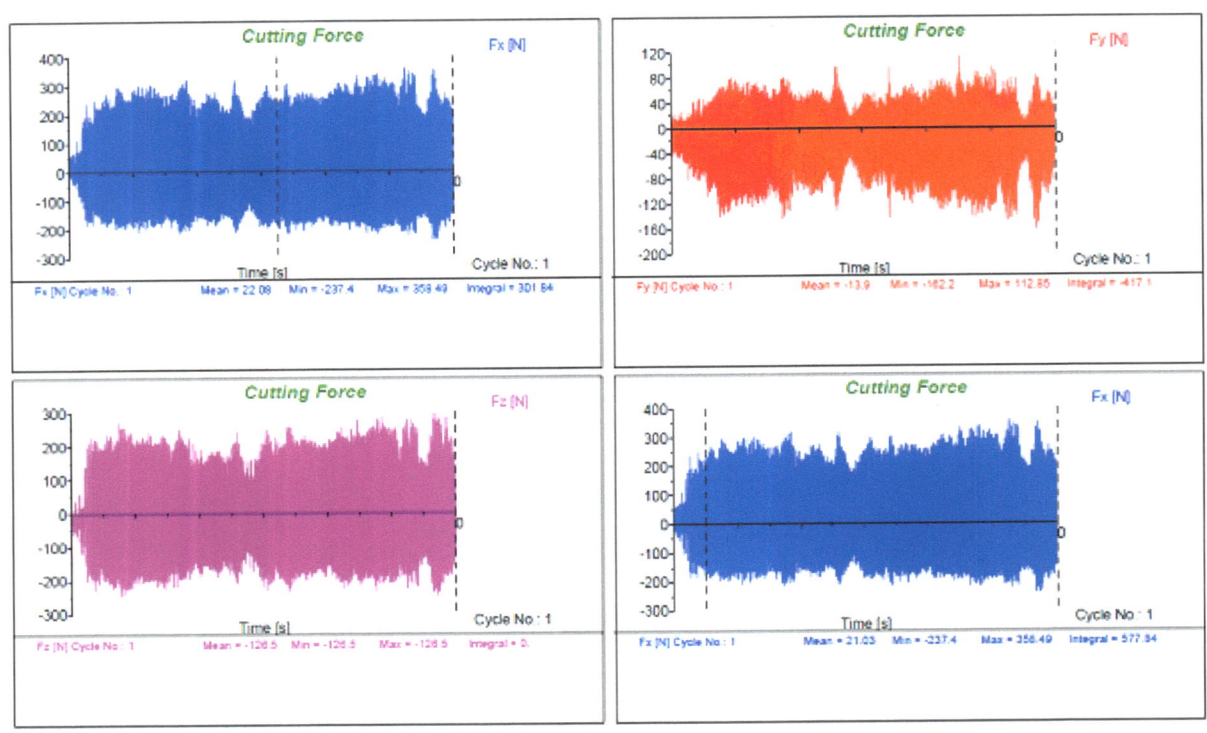

GRAPH 5.31- WITH DAMPER-CASTIRON FOR 0.02mm DEPTH, 500rpm

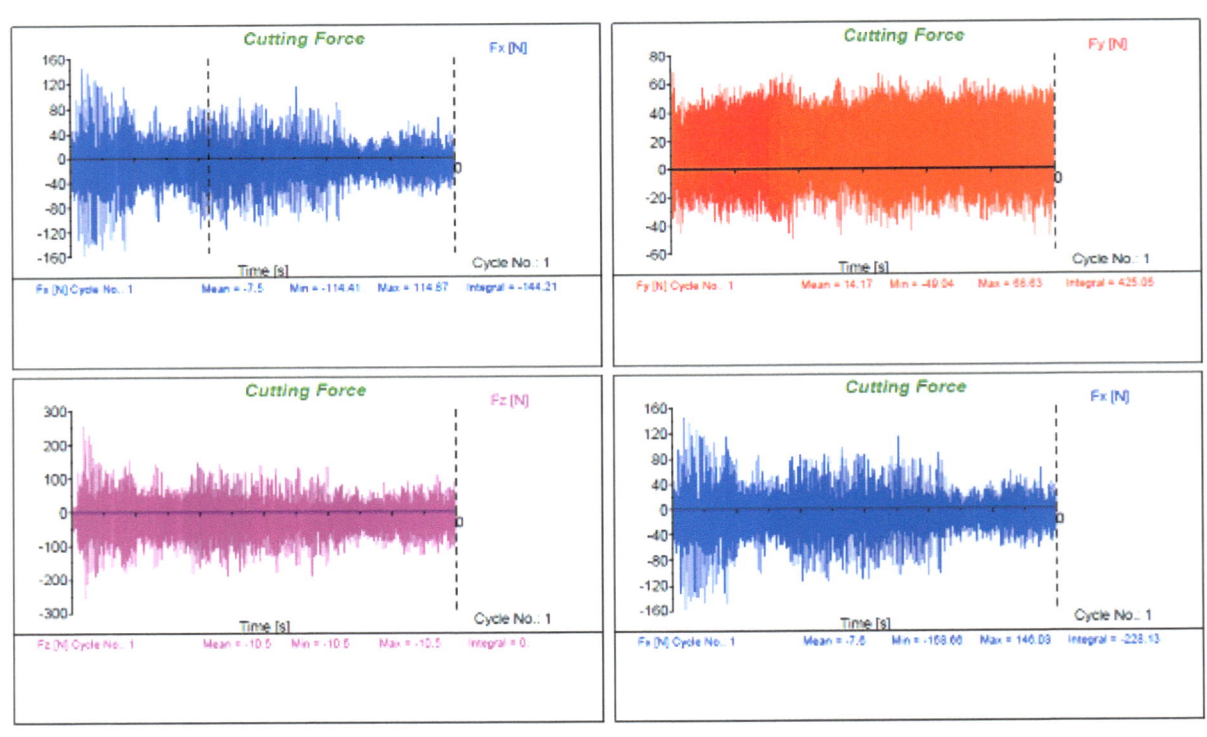

GRAPH5.32-WITHDAMPER-CASTIRON FOR0.02mm DEPTH,600mm

CUTTING FORCE GRAPH FOR THE BORING TOOL WITH DAMPING MATERIAL OF COPPER

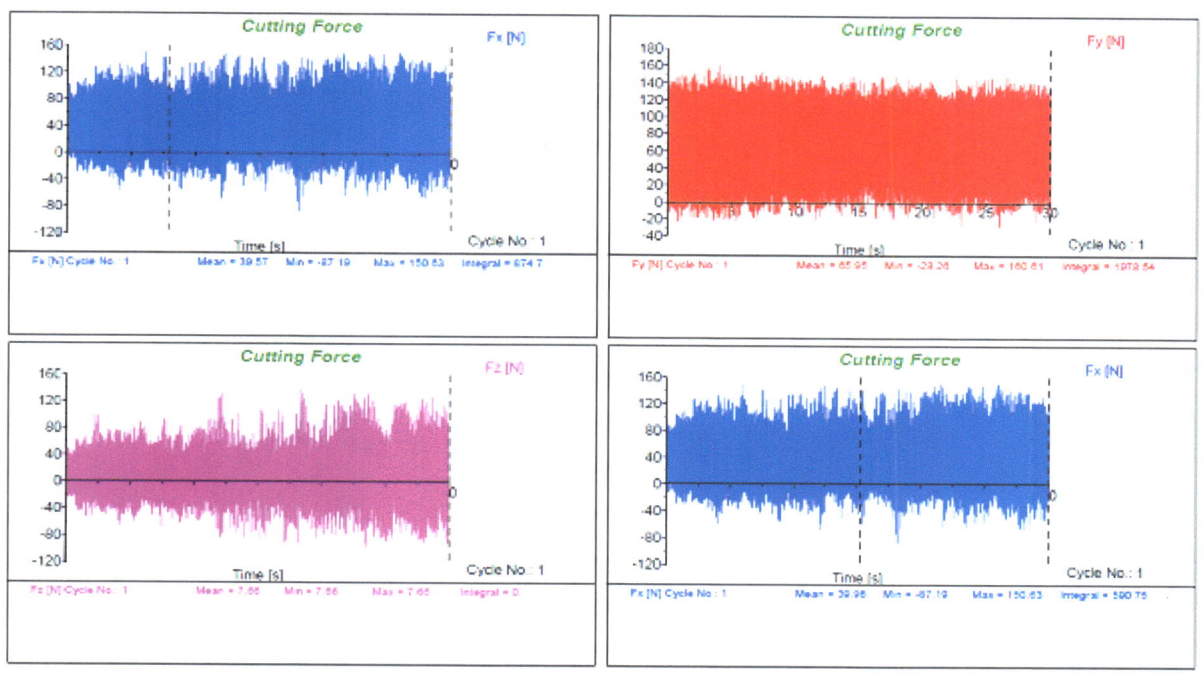

GRAPH 5.33-WITH DAMPER-COPPER FOR 0.01mm DEPTH, 500rpm

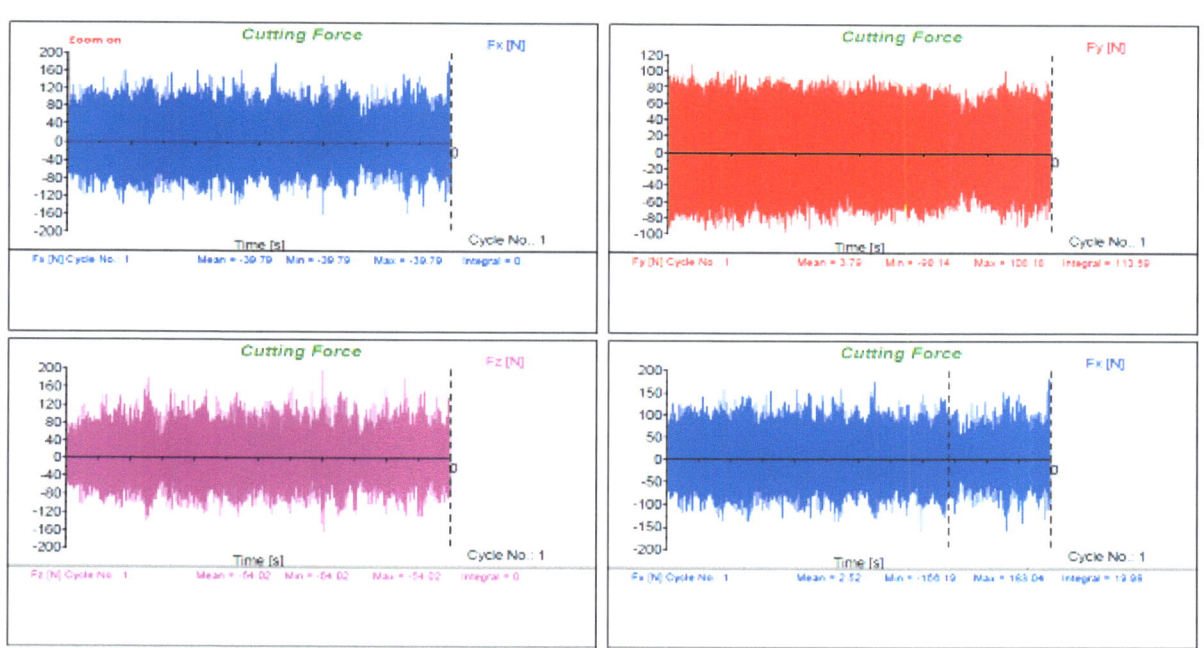

GRAPH 5.34-WITH DAMPER –COPPER FOR0.01mm DEPTH, 600rpm

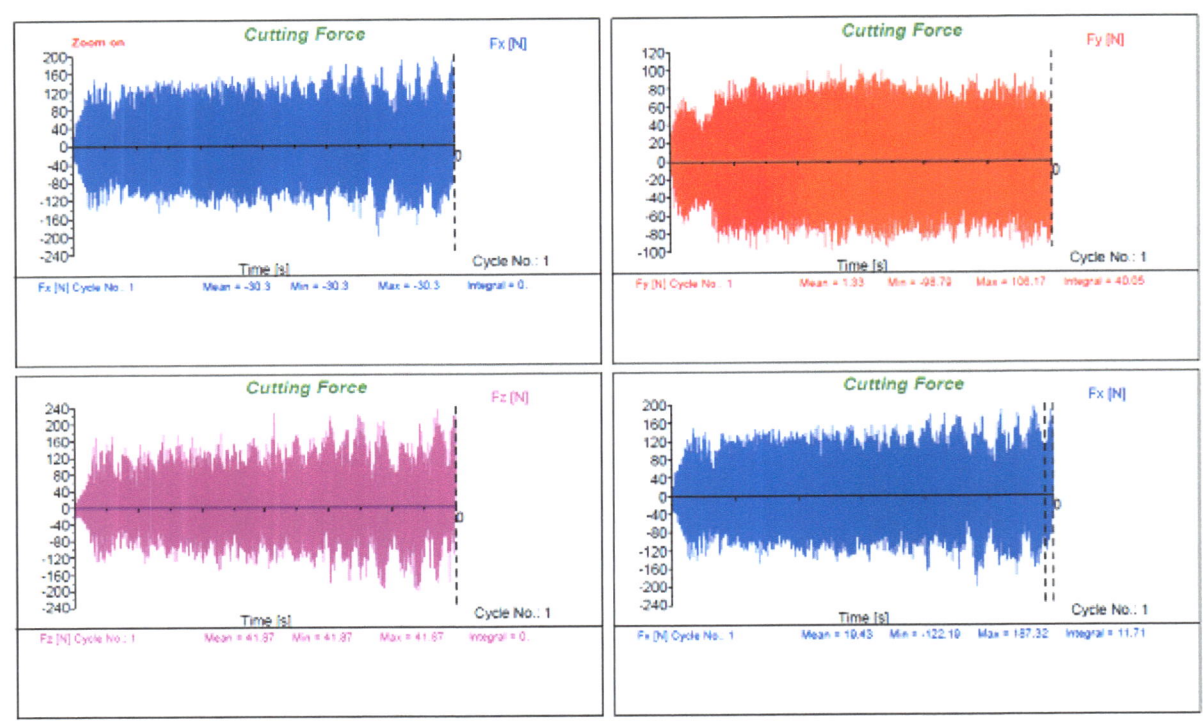

GRAPH5.35-WITH DAMPER-COPPER FOR 0.02mm DEPTH, 500rpm

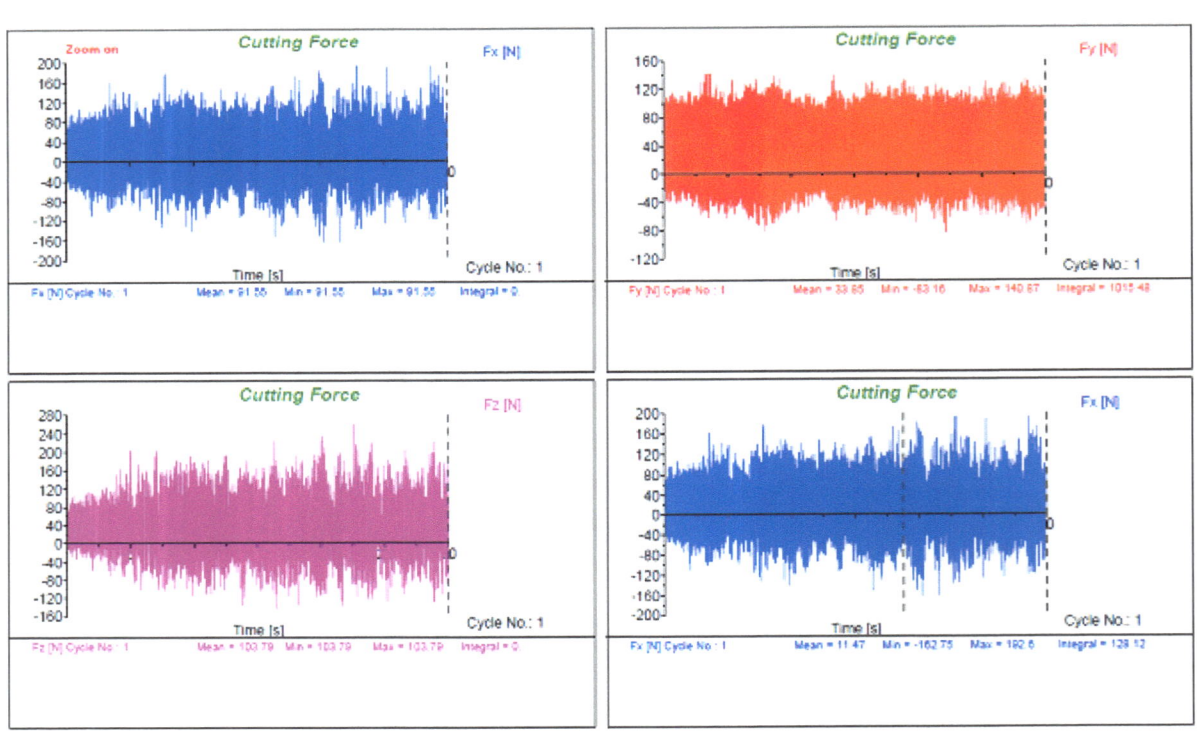

GRAPH5.36-WITH DAMPER-COPPER FOR 0.02mm DEPTH,600rpm

CUTTING FORCE GRAPH FOR THE BORING TOOL WITH DAMPING MATERIAL OF EN31

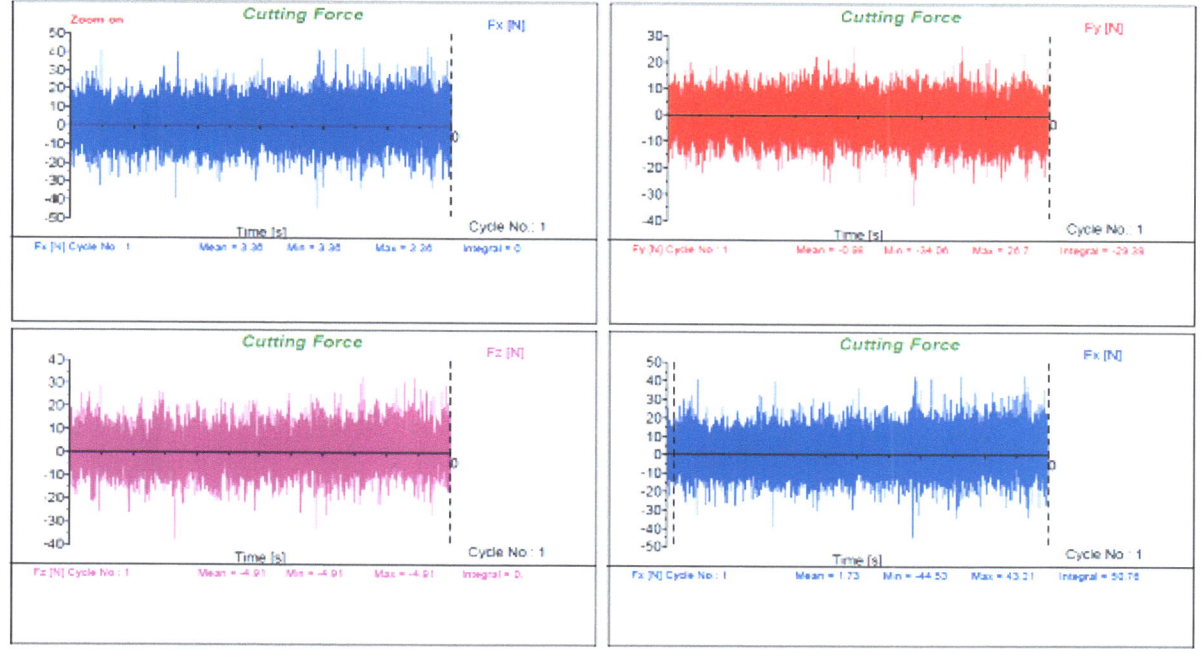

GRAPH 5.37- WITH DAMPER-EN31 FOR0.01mm DEPTH, 500rpm

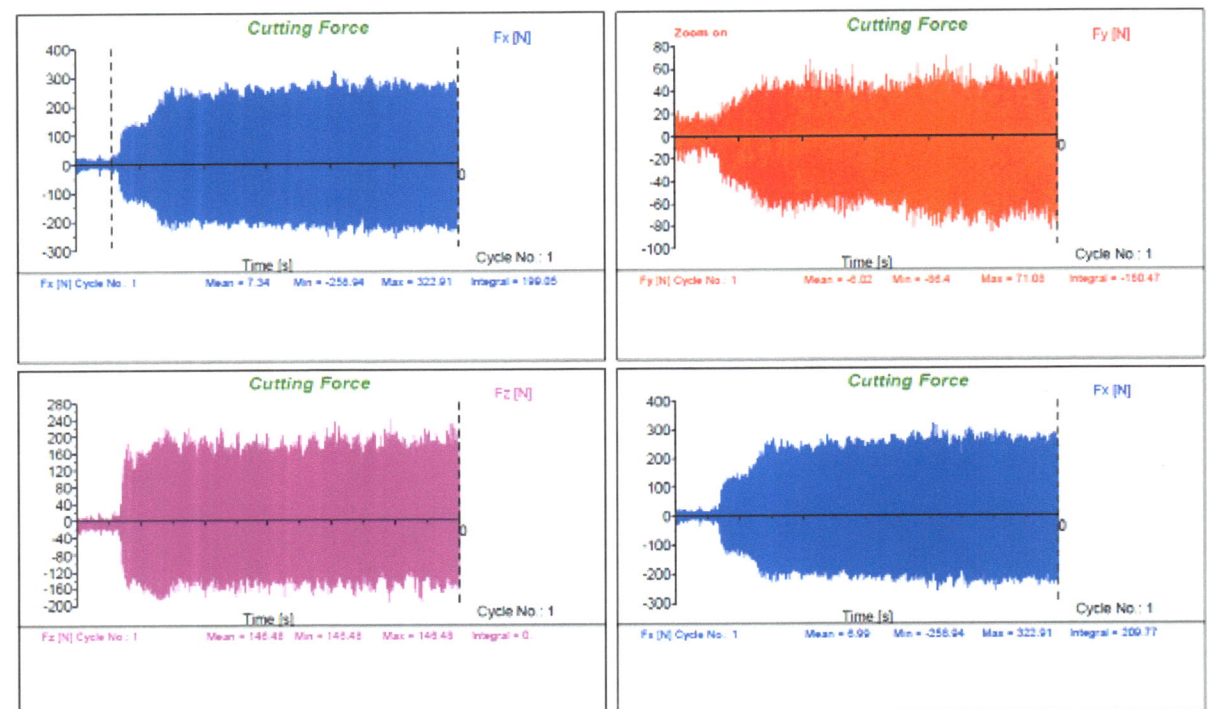

GRAPH 5.38- WITH DAMPER-EN31 FOR 0.01mm DEPTH, 600rpm

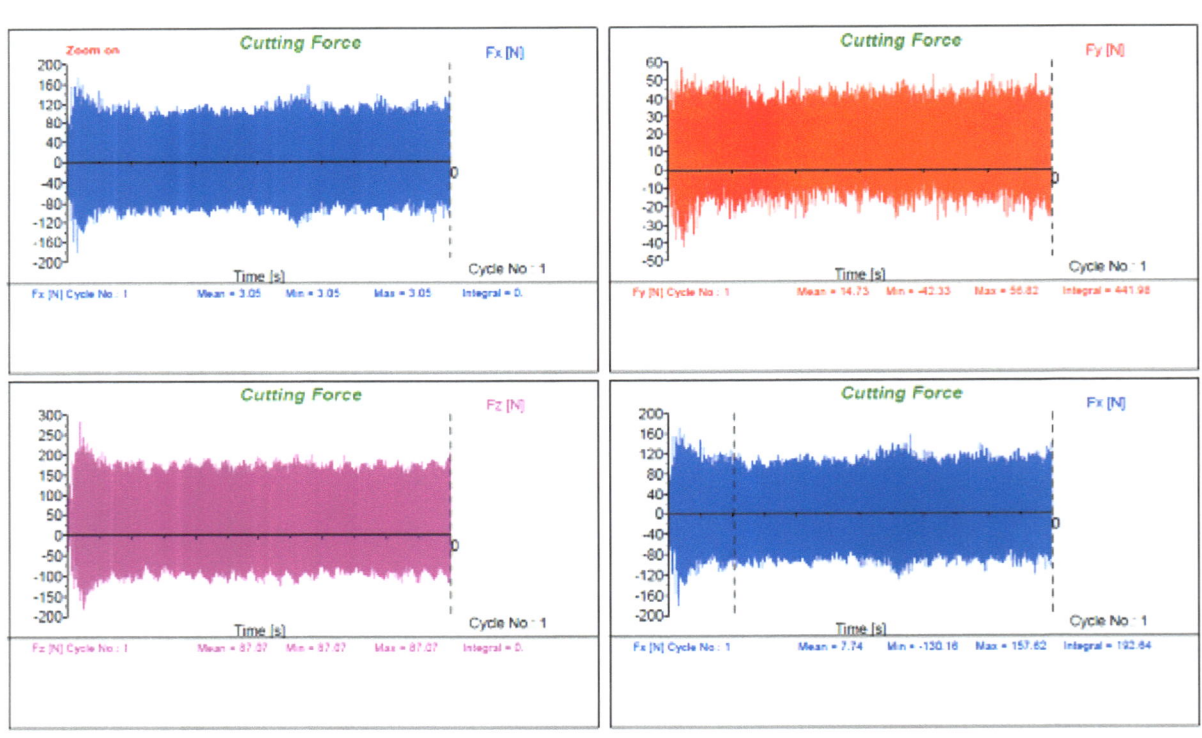

G RAPH -5.39 WITH DAMPER-EN31 FOR0.02mm DEPTH, 500rpm

CHAPTER 6
INVESTIGATION OF THERMAL BEHAVIOUR OF BORING TOOLS
USING ANSYS

6.1 MESHED MODEL WITHOUT BORING TOOL

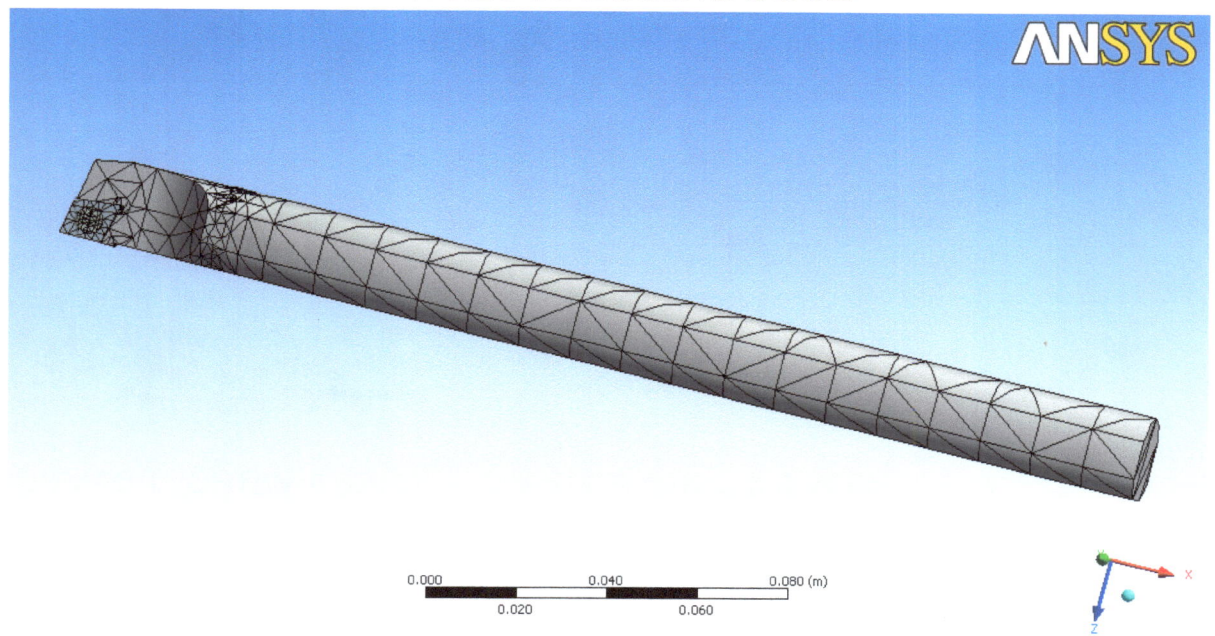

FIG 6.1-MESHED MODEL WITHOUT BORING TOOL

6.2 MESHED MODEL WITH DAMPING BORING TOOL

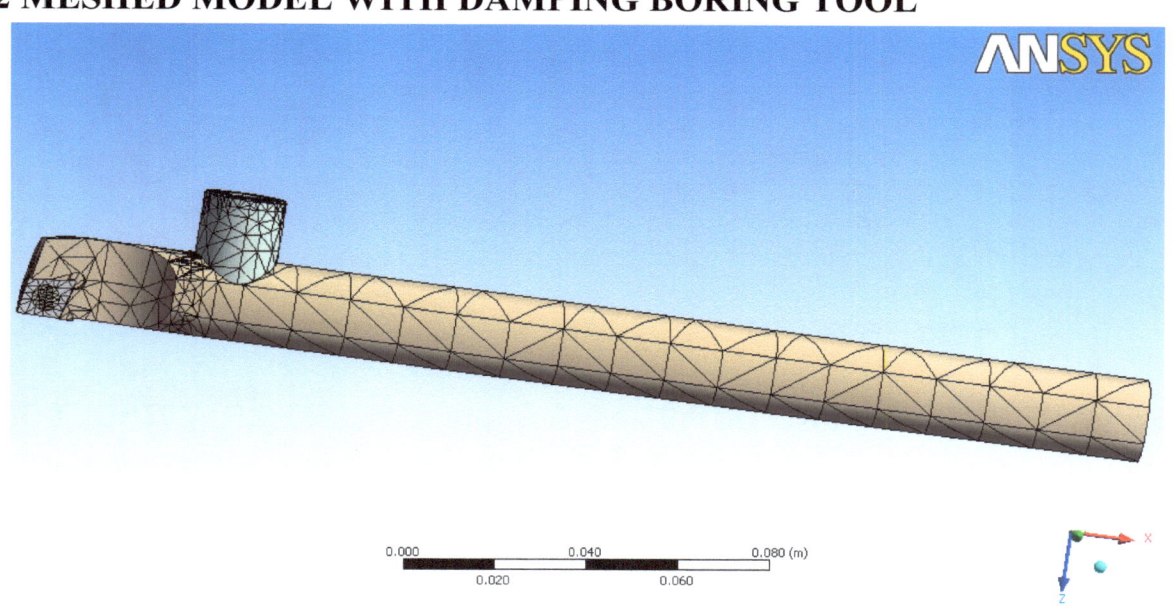

FIG6.2-MESHED MODEL WITH DAMPING BORING TOOL

6.3 TEMPERATURE DISTRIBUTION OF WITHOUT DAMPING BORING TOOL

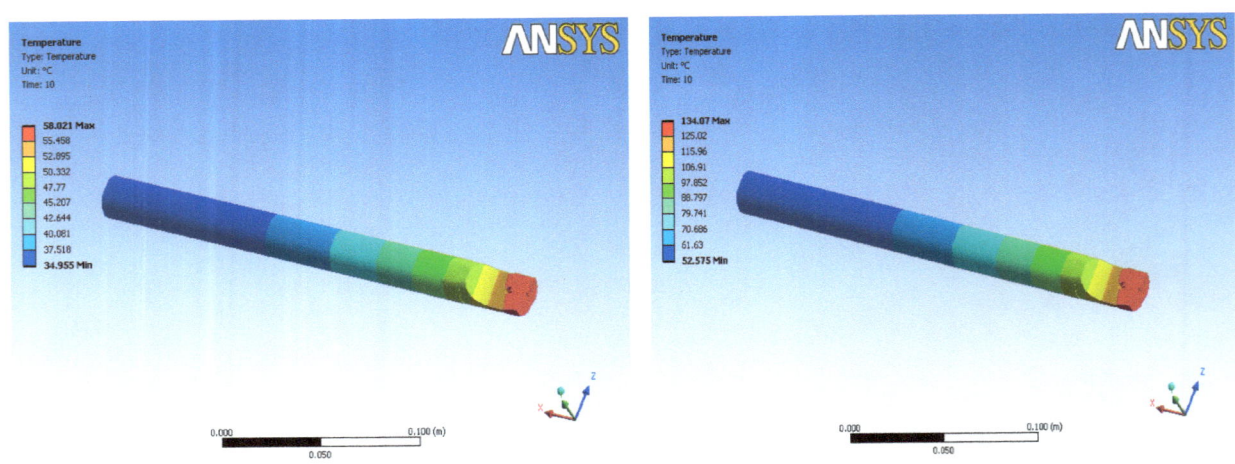

FIG 6.3-WITHOUT DAMPING FOR 0.01mm, 500 rpm 0.02mm,500 rpm

FIG6.4-WITHOUT DAMPING FOR

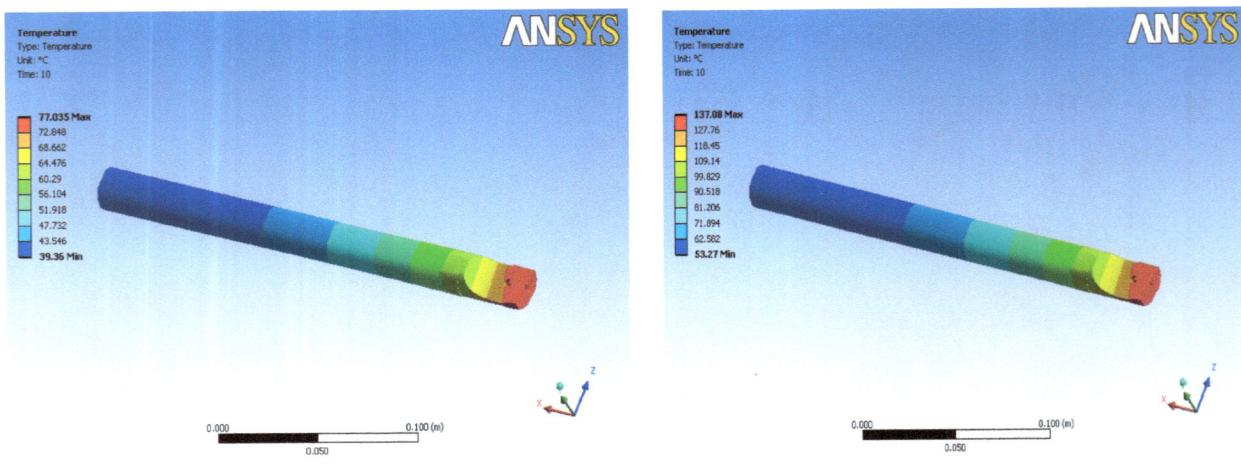

FIG 6.5 –WITHOUT DAMPING FOR 0.01mm,600 rpm 0.02mm,600 rpm

FIG 6.6 –WITHOUT DAMPING FO

6.4 HEAT FLUX DISTRIBUTION OF WITHOUT DAMPING BORING TOOL

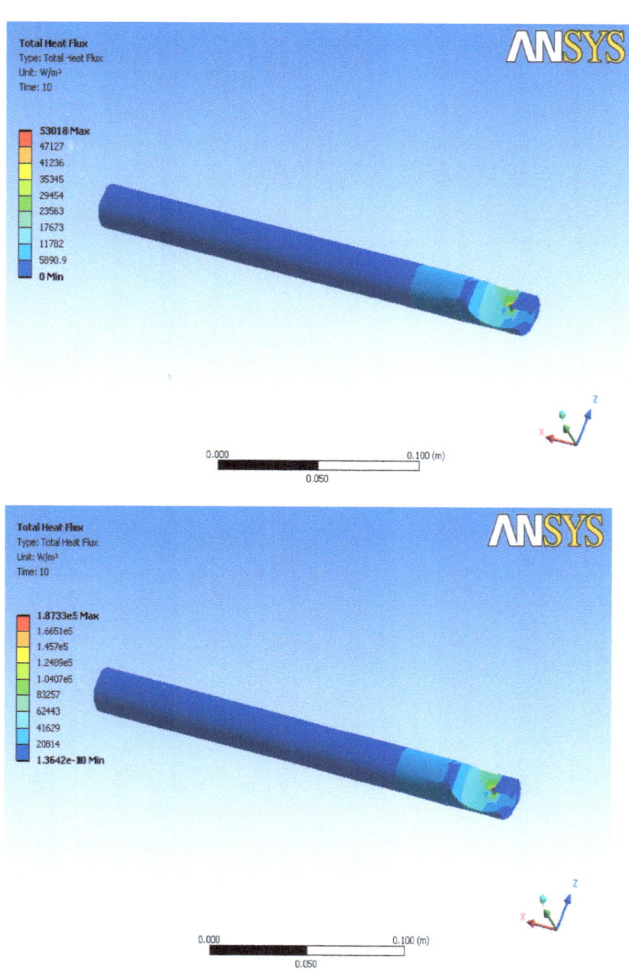

FIG 6.7 –WITHOUT DAMPING, 0.01mm,500 rpm   FIG 6.8 –WITHOUT DAMPING, 0.02mm,500 rpm

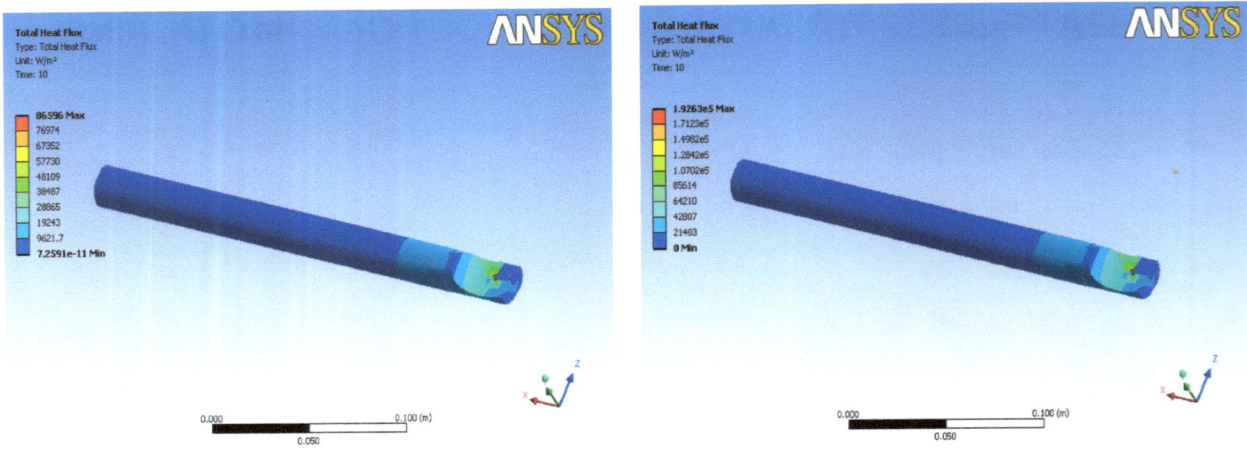

FIG 6.9 –WITHOUT DAMPING, 0.01mm,600 rpm   FIG 6.10 –WITHOUT
DAMPING,0.02mm 600 rpm

6.5 TEMPERATURE DISTRIBUTION IN THE BORING TOOL WITH DAMPING MATERIAL OF BRASS

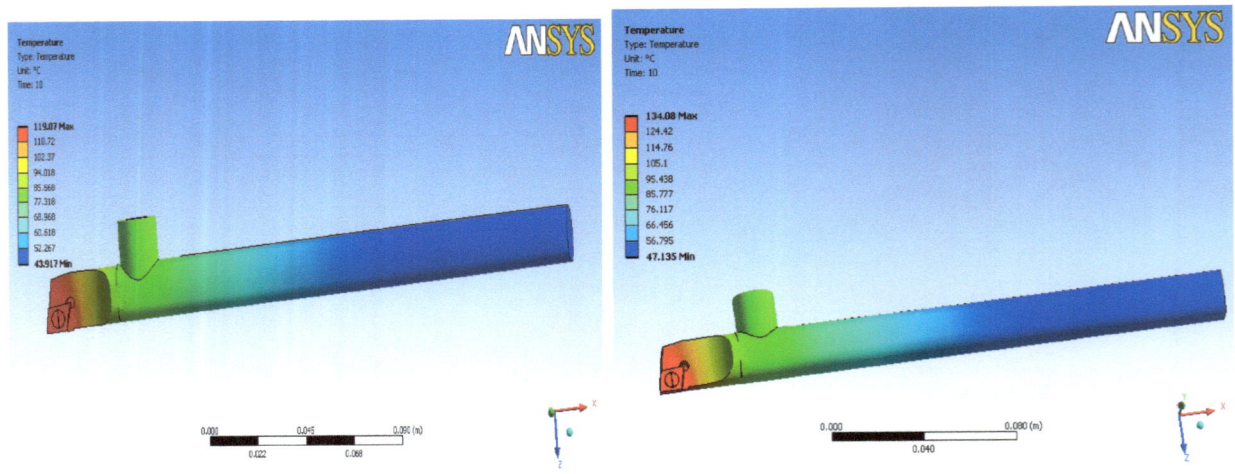

FIG 6.11 –WITH DAMPER-BRASS,0.01mm,500 rpm FIG 6.12 – WITH
DAMPER-BRASS, 0.02mm, 500rpm

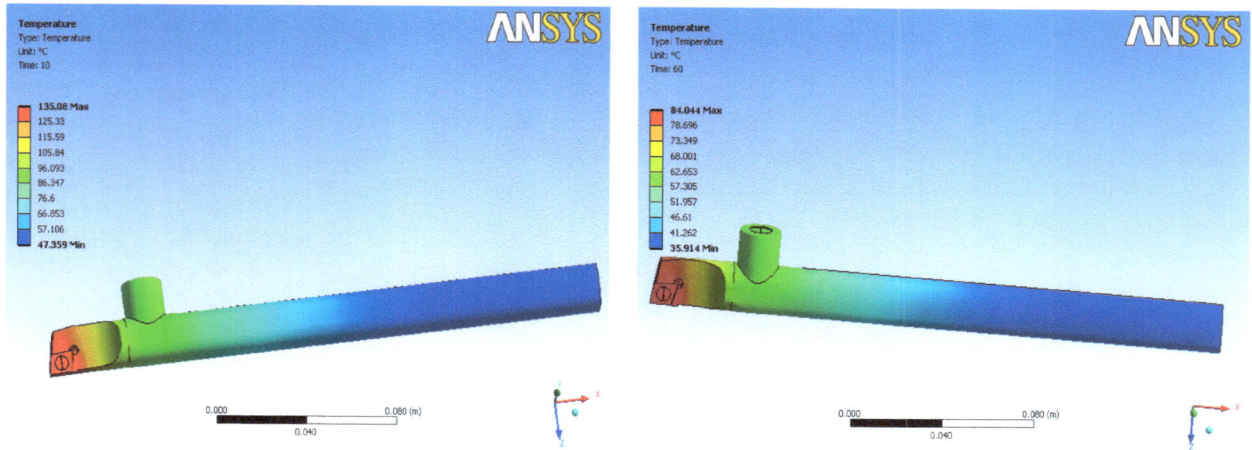

FIG 6.13 –WITH DAMPER-BRASS, 0.01mm 600 rpm   FIG 6.14 –WITH DAMPER-BRASS, 0.02mm 600 rpm

6.6 HEAT FLUX DISTRIBUTION IN THE BORING TOOL WITH DAMPING MATERIAL OF BRASS

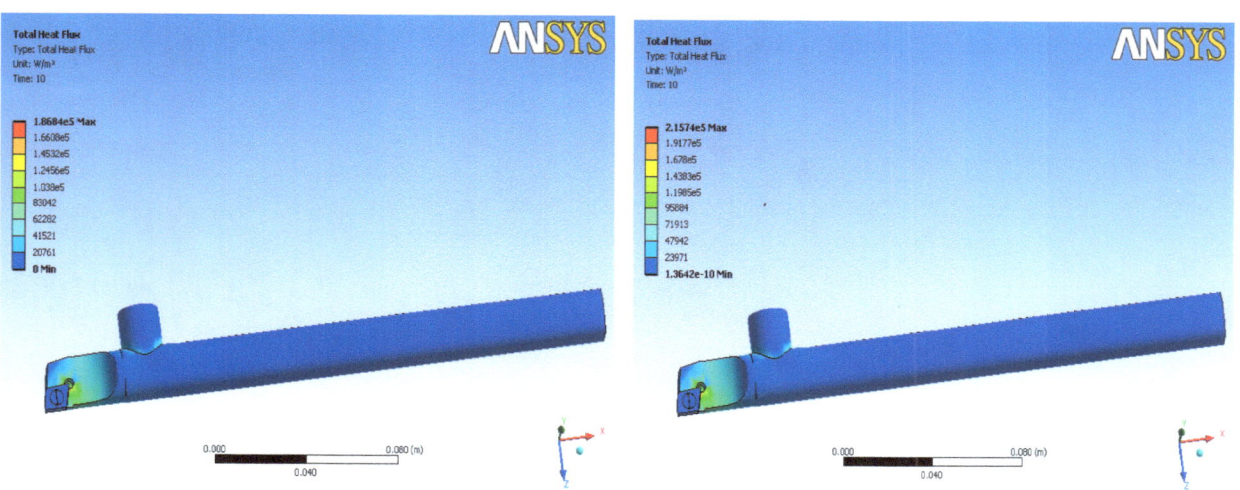

FIG   6.15   –WITH   DAMPER-BRASS0.01mm,500   rpm   FIG   6.16   –WITH DAMPER-BRASS, 0.02mm,500rpm

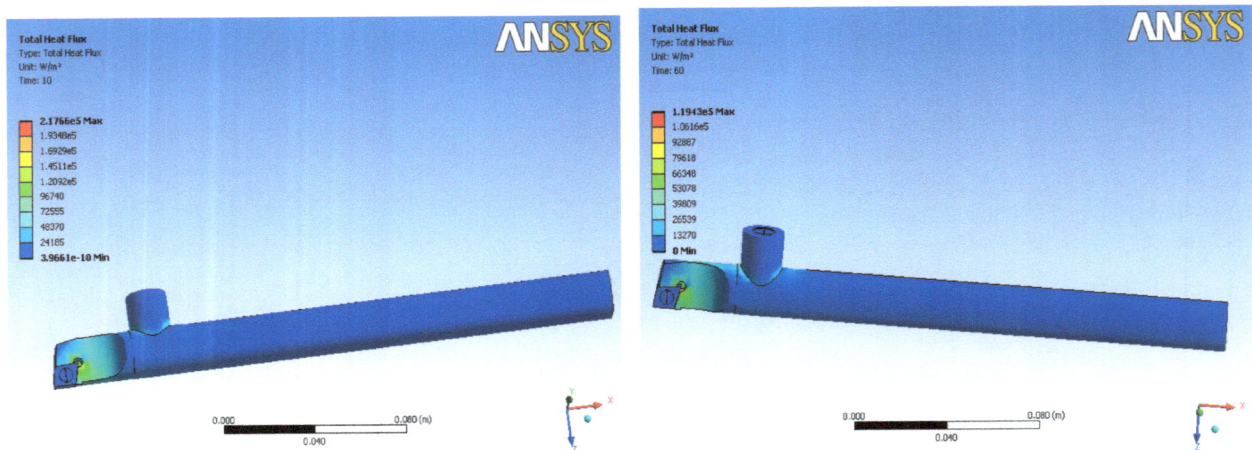

FIG 6.17 –WITH DAMPER-BRASS,0.01mm, 600rpm FIG 6.18 –WITH DAMPER-BRASS,0.02mm,600 rpm

6.7 TEMPERATURE DISTRIBUTION IN THE BORING TOOL WITH DAMPING MATERIAL OF CASTIRON

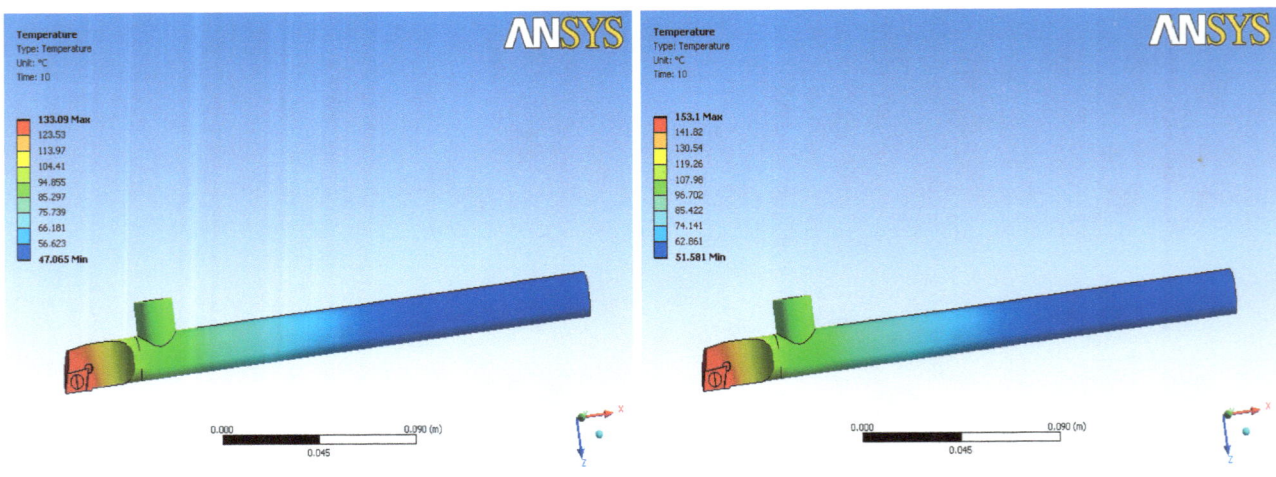

FIG 6.19 –WITH DAMPER-CASTIRON,0.01mm, 500 rpm   FIG 6.20 –WITH DAMPER-CASTIRON, 0.02mm,500 rpm

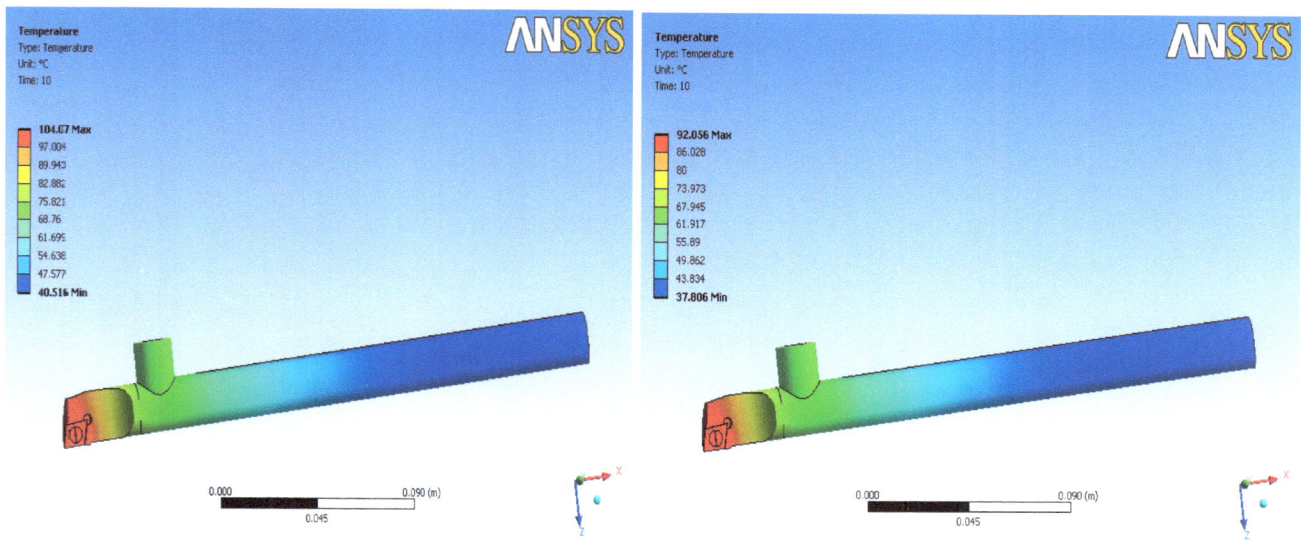

FIG 6.21 –WITH DAMPER-CASTIRON, 0.01mm,600 rpm   FIG 6.22 –WITH DAMPER-CASTIRON,0.02mm ,600 rpm

6.8 HEAT FLUX DISTRIBUTION IN THE BORING TOOL WITH DAMPING MATERIAL OF CASTIRON

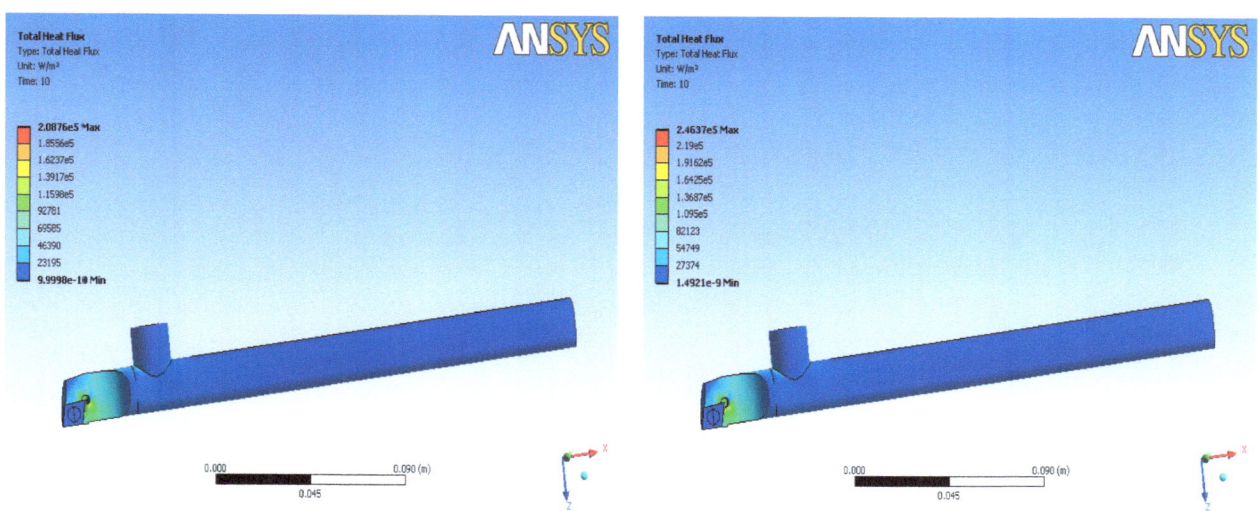

FIG 6.23 –WITH DAMPER-CASTIRON,0.01mm, 500 rpm FIG 6.24 –WITH DAMPER-CASTIRON, 0.02mm, 500 rpm

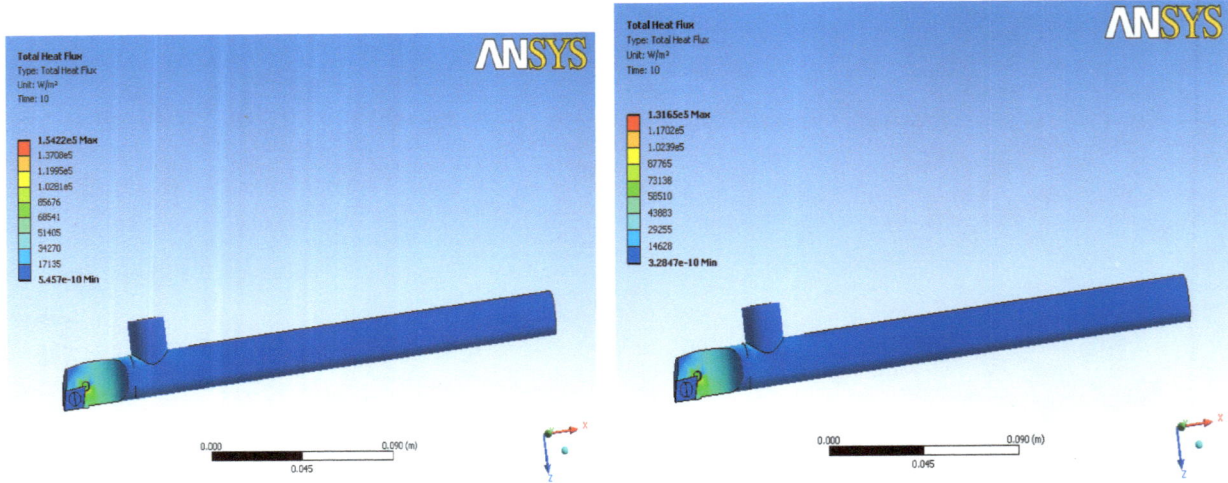

FIG 6.25 –WITH DAMPER-CASTIRON, 0.01mm, 600rpm   FIG 6.26 –WITH DAMPER

CASTIRON, 0.02mm,600 rpm

6.9 TEMPERATURE DISTRIBUTION IN THE BORING TOOL WITH DAMPING MATERIAL OF EN31

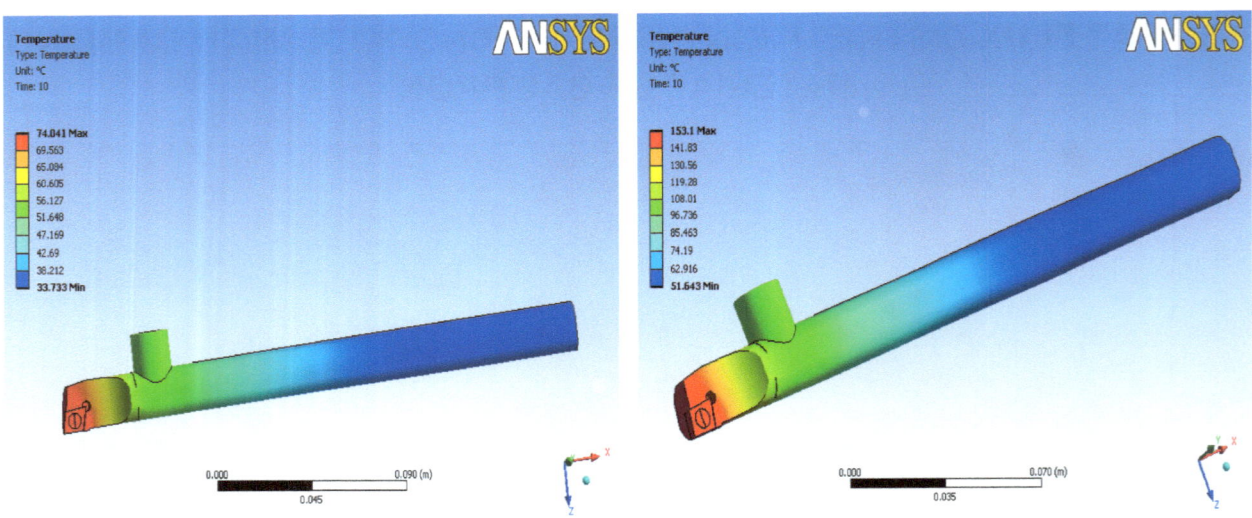

FIG 6.27 –WITH DAMPER-EN31,0.01mm, 500 rpm   FIG 6.28 –WITH

DAMPER-EN31, 0.02mm, 500 rpm

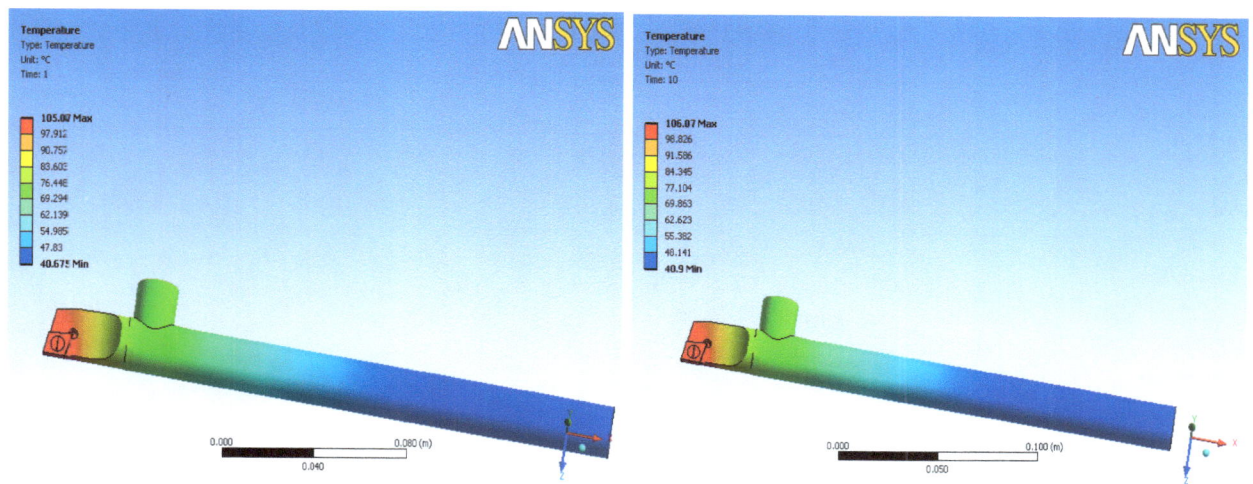

FIG 6.29 –WITH DAMPER-EN31, 0.01mm, 600 rpm   FIG 6.30 –WITH DAMPER-EN31, 0.02mm, 600 rpm

6.10 HEAT FLUX DISTRIBUTION IN THE BORING TOOL WITH DAMPING MATERIAL OF EN31

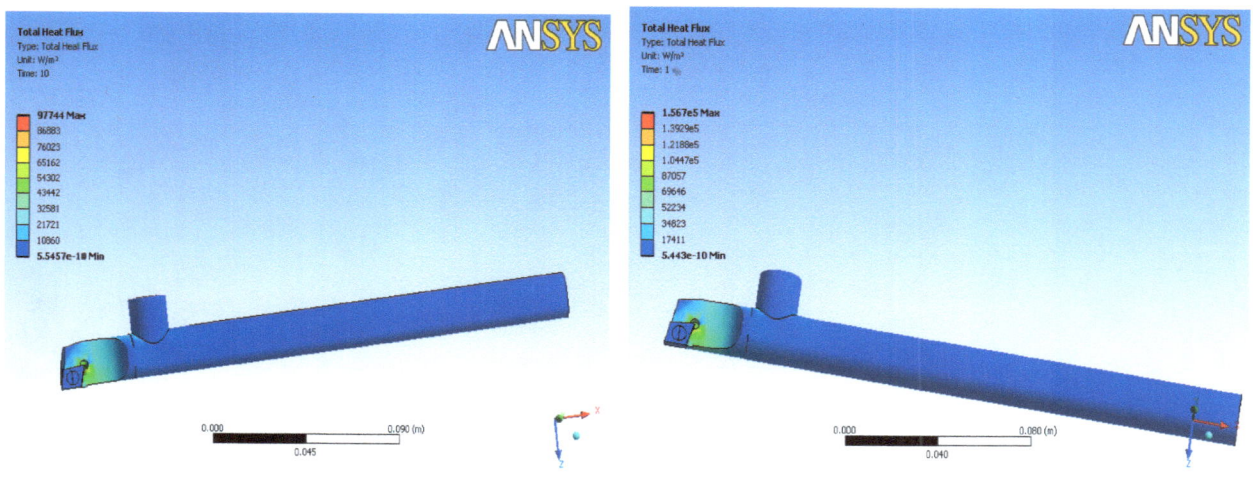

FIG 6.31 –WITH DAMPER-EN31 FOR 0.01mm, 500 rpm FIG 6.32 –WITH DAMPER-EN31 FOR 0.02mm, 500 rpm

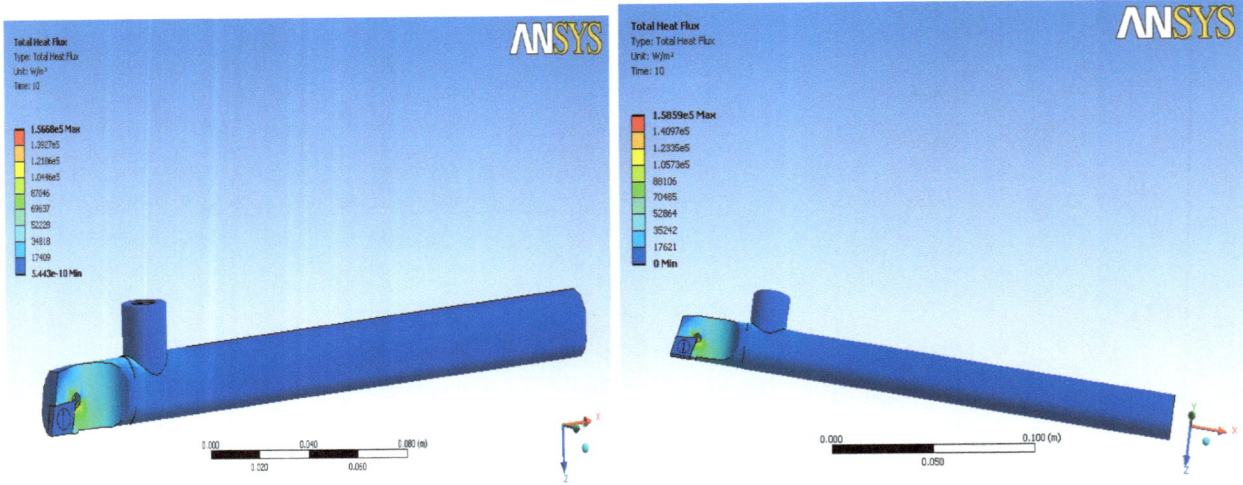

FIG 6.33 –WITH DAMPER-EN31, 0.01mm, 600 rpm   FIG 6.34 –WITH
DAMPER-EN31, 0.02mm, 600 rpm

6.11 TEMPERATURE DISTRIBUTION IN THE BORING TOOL WITH DAMPING MATERIAL OF COPPER

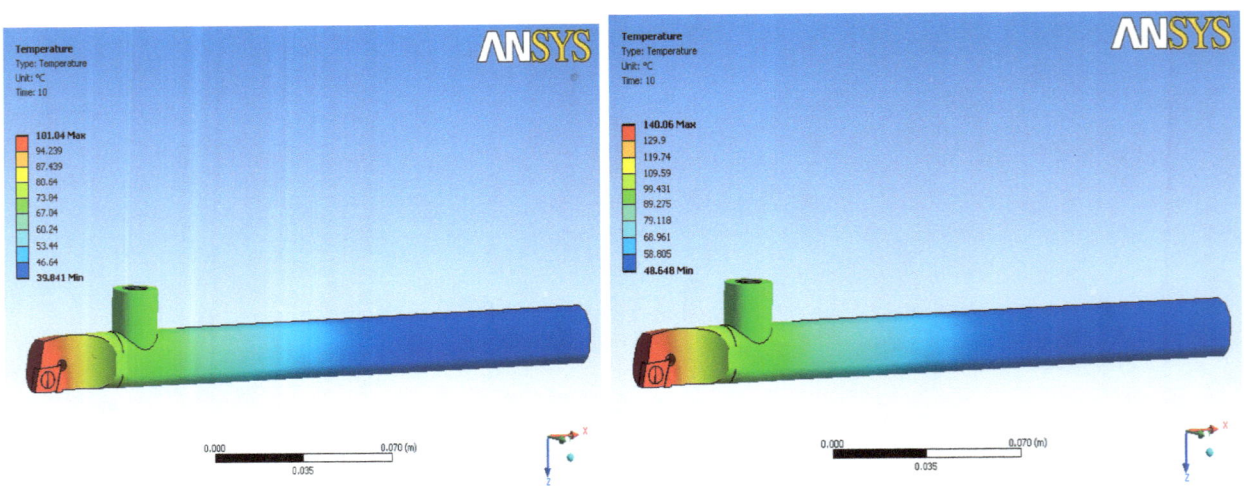

FIG 6.35 –WITH DAMPER-Cu, 0.01mm,500 rpm   FIG 6.36 –WITH
DAMPER-Cu ,0.02mm,500 rpm

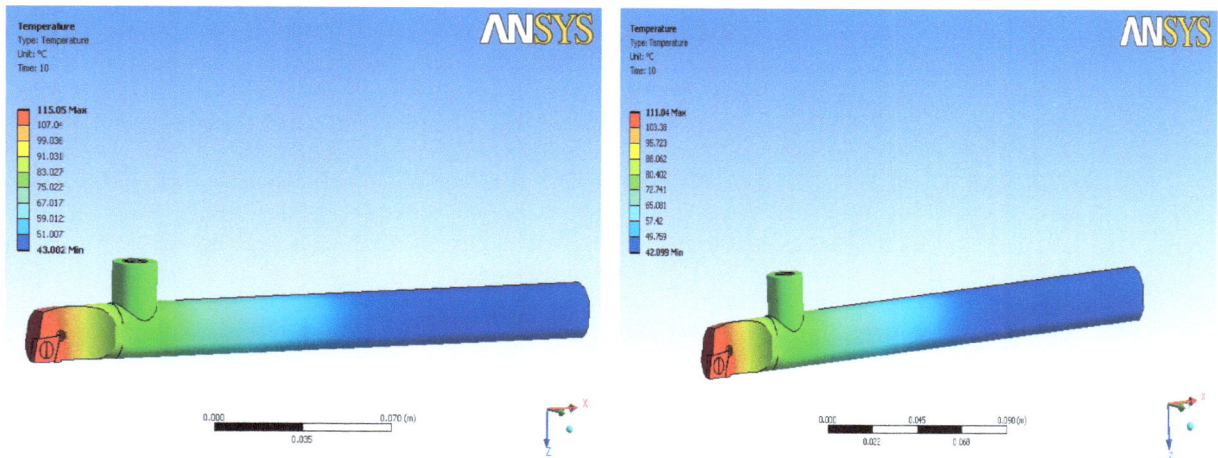

FIG 6.37 –WITH DAMPER-COPPER,0.01mm, 600 rpm FIG 6.38 –WITH DAMPER-COPPER,0.02mm,600 rpm

6.12 HEAT FLUX DISTRIBUTION IN THE BORING TOOL WITH DAMPING MATERIAL OF COPPER

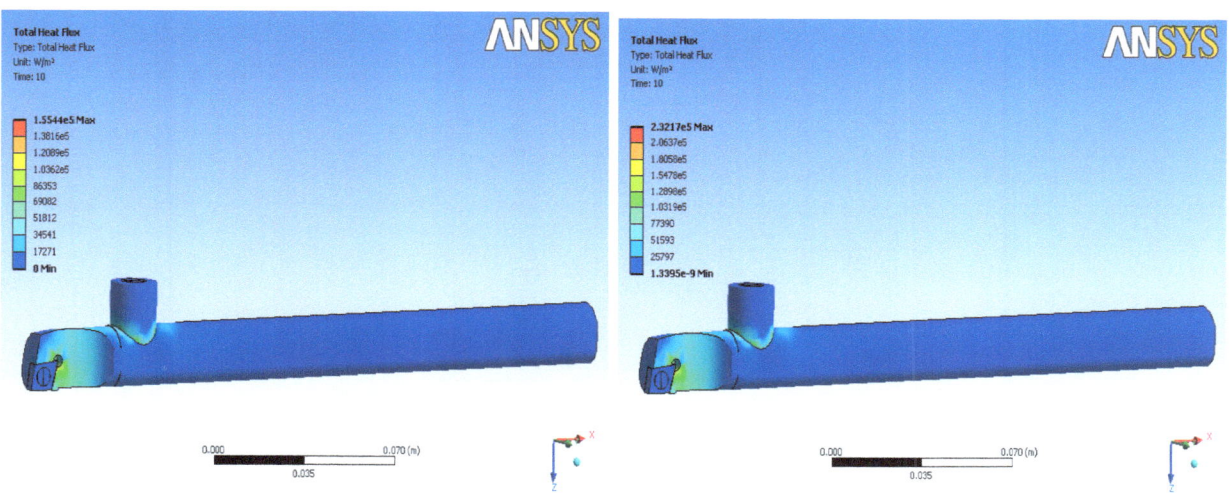

FIG 6.39 –WITH DAMPER-COPPER,0.01mm, 500 rpm   FIG 6.40 –WITH DAMPER-COPPER,0.02mm ,500 rpm

CHAPTER 7

RESULTS AND DISCUSSION

By using various damping materials, the tests were conducted and from the graph, the parameters (cutting force and temperature of the insert) are noted down and tabulated.

Damping material	Speed in rpm		Depth in mm			
			0.01	0.02	0.01	0.02
			Cutting force in N		Temperature in °C	
Without damper	500	F_X	118.3	137.08	58	134
		F_Y	35.31	33.26		
		F_Z	84.14	225.4		
	600	F_X	37.51	134.19	76.6	137
		F_Y	25.45	56.46		
		F_Z	48.13	378		
Copper	500	F_X	150.63	195.53	101	140
		F_Y	160.61	106.17		
		F_Z	139.13	235.25		
	600	F_X	176.39	192.6	115	111
		F_Y	108.18	140.6		
		F_Z	197.33	258.79		
Brass	500	F_X	87.8	129.64	118.5	133.7
		F_Y	51.54	42.60		
		F_Z	164.55	191.71		
	600	F_X	160.13	151.37	135	83.5
		F_Y	113.77	95.18		
		F_Z	232.15	175.81		
Cast Iron	500	F_X	289.58	358.49	133	152.5
		F_Y	45.72	112.85		
		F_Z	246.77	395.38		
	600	F_X	44.16	137.57	104	92.5
		F_Y	58.65	67.75		
		F_Z	38.12	254.52		
	500	F_X	43.21	160.06	74	152.8

EN31		F$_Y$	26.7	56.82		
		F$_Z$	32.96	244.23		
	600	F$_X$	322.91	339.08	105	105.58
		F$_Y$	71.08	7.78		
		F$_Z$	242.55	285.092		

TABLE 7.1- TEMPERATURE READINGS FOR TESTS

The vibration results for various depths and speeds, are also measured by vibrometer.

Damping Material	SPEED in rpm	DEPTH in mm			
		0.01mm		0.02mm	
		ACCELERATION	DISPLACEMENT	ACCELERATION	DISPLACEMENT
COPPER	500	257.57	0.08	227.5	0.0725
	600	333.33	0.0625	291.75	0.4325
BRASS	500	183.62	0.0525	236	0.0838
	600	260.85	0.061	245.75	0.1075
CAST IRON	500	228.83	0.3	360	0.02
	600	99	0.0475	272.25	0.085
EN31	500	35.71	0.01625	120.25	0.04
	600	291.33	0.525	356.5	0.7025
WITHO	500	59	0.015	160	0.27

UT DAMPER	600	51	0.0347	156	0.9075

TABLE -7.2 VIBRATION RESULTS FROM THE EXPERIMENTS

From the above table , vibration produced in tool for various depths and speeds, compare the values the copper is lesser vibration and produced than the other damping materials.

From the ansys results, compare the heat flux distribution simulations, the copper damped tool is better than the other damped tools like cast iron ,brass, and structured steel(EN31).

CHAPTER 8

CONCLUSION

From the experimental results obtained by using various damping materials, (copper, cast iron, EN31 and brass) vibration and temperature produced in the copper damped tool was lesser than the other damped tools. Moreover, from the ANSYS results, it is found out that heat transfer is very high when copper is used as damping material. Hence, it is suggested that copper can be used as a damping material for boring tool.

BIBLIOGRAPHY

1. C.Mei, 2004, "Active regenerative chatter suppression during boring manufacturing process".

2. A.Ganguli, A.Deraemaeker, A.Preumont, 2007 "Regenerative chatter reduction by active damping control", Journal of sound and vibration 300, 847-862.

3. N.D. Sims and Y. Zhang, Amas, 2003,"Active damping for chatter reduction in high speed machining", Workshop on smart materials and structures smart'03, pp.195–212.

4. L. Andren, I. H.akansson, i. Claesson, 2005, "Actuator placements and variations in the control path estimates in the active control of boring bar vibrations", department of signal processing, Blekinge Institute Of technology, Sweden.

5. Lonnie houck, tony l.schmitz, and k.scott smith, 2006," Tunable Holder for boring bars", University of Floride, USA.

6. Eugene .I. Rivin, 2007, "Use of Stiffness, damping, natural frequency criteria in vibration control", Int.J. of Machine Tools and Manufacture,vol.32.

7. Dongki won, 2004, "Numerical analysis and parameter study of a mechanical damper in machine tool", university of florida, USA.

www.ingramcontent.com/pod-product-compliance
Lightning Source LLC
Chambersburg PA
CBHW050814180526
45159CB00004B/1660